Sea-Level Change in the Gulf of Mexico

TEXAS A&M
UNIVERSITY
CORPUS
CHRISTI

HARTE
RESEARCH INSTITUTE
FOR GULF OF MEXICO STUDIES

John W. Tunnell Jr., General Editor

Texas A&M
University Press
College Station

Sea-Level Change in the
Gulf of Mexico

Richard A. Davis Jr.

Copyright © 2011 by Richard A. Davis Jr.

Printed in China by Everbest Printing Co., through FCI Print Group

First edition

This paper meets the requirements of ANSI/NISO Z39.48-1992
(Permanence of Paper).
Binding materials have been chosen for durability.

Library of Congress Cataloging-in-Publication Data
Davis, Richard A. (Richard Albert), 1937–
Sea-level change in the Gulf of Mexico / Richard A. Davis Jr.—1st ed.
        p. cm.—(Harte Research Institute for Gulf of Mexico Studies Series)
Includes bibliographical references and index.
ISBN 978-1-60344-224-4 (flexibound : alk. paper) 1. Sea level—Mexico, Gulf of. 2. Coast
changes—Mexico, Gulf of. 3. Mexico, Gulf of. I. Title. II. Series: Harte Research Institute for Gulf
of Mexico Studies Series.
 GC90.M62D38 2011
 551.45'80916364—dc22
                                            2010035648

Contents

Preface

People in the early civilizations recognized that sea level has changed over time. Stories in ancient Greek and Hebrew described great floods of the sea. The presence of fossils and huge boulders well above present sea level were used as evidence. Much later, in the seventeenth century, Neptunism was prevalent among scientists. This theory held that there was a major fall in sea level. As more data were collected and naturalists became more sophisticated in their interpretation, it was recognized that phenomena such as earthquakes and volcanic eruptions could cause sea-level change. Finally, in the late nineteenth century, the famous scientist Eduard Suess recognized that sea level has changed globally; he termed this "eustatic" change. That is, sea level can change throughout the world, both upward and downward.

The Gulf of Mexico is a mediterranean sea, a large water body surrounded by land. The Mediterranean Sea is an excellent example as are the Persian Gulf and the Red Sea. All of these large water bodies owe their origin to plate tectonics. How old is the Gulf of Mexico, how did it come about, and how has sea-level change influenced its present appearance? These and other topics will be discussed in this book.

Humans populate all the coasts of the world, and changes in sea level that are experienced by coasts impact the nearby population. The topic of global warming, which can produce sea-level change, especially sea-level rise, is now in the media almost daily. As little as a couple of decades ago, people had no concern about global warming at all.

This book is not aimed at a discussion of the causes of global warming but only at sea-level change, both its rise and its fall. There are many factors that can contribute to this condition, some of which are due to climate change and others that are not. All of these factors will be considered.

The Gulf of Mexico coast is densely populated in the United States but much less so in Mexico and Cuba. More than 25 million people live along the Gulf, including large metropolitan areas around Houston in Texas and the Tampa Bay area in Florida. There are also extensive coastal reaches with little or no population including much of the Texas Gulf Coast and nearly all the Mexico Gulf Coast. The economy of the Gulf of Mexico coast is based largely on the petroleum industry, tourism, and shipping with a diminishing seafood industry a distant fourth. Sea-level change has and will have, a significant impact on these industries and therefore on the economy of the Gulf Coast.

The contents of this book will provide the reader with a comprehensive discussion of sea-level change, its causes, types and rates, the geologic history of sea-level change, our present situation, and some potential effects in the future. Sea-level change is largely based on geological phenomena. Knowledge of that discipline is not required for a thorough understanding of this text. This book is written for the general public including coastal managers, government officials, coastal engineers, biologists, and others with an interest in this important and timely topic. Both the text and illustrations come from a wide range of sources; little is original with the author. The book is a synthesis of what we now know about sea-level change along the margins of this important body of water. The numerous references cited in books for research purposes are eliminated in favor of a brief list of related readings. Specific sources in the scientific literature are provided in the figure captions indicating their origin.

I would be remiss without mentioning the several outstanding research efforts and subsequent publications on the Gulf of Mexico from various individuals and groups. The U.S. Geological Survey, the Florida Geological Survey and The University of Texas—Bureau of Economic Geology have led prominent efforts. Ervin Otvos of the Gulf Springs Research Laboratory in Mississippi has been a prolific researcher on the northern Gulf of Mexico coast. The Coastal Studies Institute of Louisiana State University has long been the dominant research group for the Mississippi River Delta, led by James Coleman and Harry Roberts. The most prominent group now working on the northern Gulf Coast down to the border with Mexico is that from the Department of Geology at Rice University led by John Anderson. He and his students have provided a great understanding of the Quaternary Period of the continental shelf and adjacent coast.

The author thanks many funding agencies that supported his research over the past forty years and the numerous colleagues and students who have helped. Illustrations were enhanced and created by Anthony

Reisinger and Fabio Moretzsohn. The Harte Research Institute of Texas A&M University–Corpus Christi provided funds for their efforts. I am very grateful to Shannon Davies, editor, and her colleagues, especially Kevin Grossman, Patricia Clabaugh, and Diana Vance at Texas A&M University Press, who helped with the production of this book. Kim Withers was an excellent copy editor. Dr. John Anderson and an unknown reviewer provided technical review.

Sea-Level Change in the Gulf of Mexico

1

Causes and Rates of Sea-Level Change

Changes in sea level can occur locally along as little as hundreds of meters of coastline, regionally across hundreds or thousands of kilometers, or globally; all may be important. Global sea-level change is generally referred to as *eustatic* change. It is eustatic or global sea level that the world's population is concerned with at the present time. The rates of change in sea level that occur locally can be very rapid, only minutes for earthquakes and hours in the case of tides, whereas significant global change in sea level occurs over centuries (Table 1-1). It is also important to realize that sea-level change is caused by many different phenomena that occur at various time and spatial scales.

We must realize that changes in sea level are actually *relative* changes between the water level and the adjacent land. Changes could occur because the land actually moved up or down, or because the volume of water in the oceans increased or decreased. Generally there is a combination of factors that causes the observed change in sea level at any particular location.

Local Sea-Level Changes

Sea-level changes that take place locally are the result of movement of the earth surface relative to sea level, either up or down. Earthquakes last for only seconds to minutes but can cause relative sea level to change by meters locally. These destructive phenomena are produced by movement of the crustal plates, typically along large and well-known fault systems. Generally there is a combination of vertical and lateral movement along these unstable fault surfaces. The vertical displacement of the earth's crust produced by the Alaskan earthquake in 1964 is a good example (Fig. 1-1). Some harbors were left high and dry (a relative drop in sea level), whereas

TABLE 1-1. Mechanisms and rates of sea-level change. From Revelle, R. 1990. *Sea Level Change*. Washington, D.C.: National Research Council, National Academy of Sciences. p. 7.

Mechanism	Time Scale(yrs)	Magnitude of Change(m)
Volume changes (temperature)		
Shallow water (0–500m)	0.1–100	0–1
Deep water (>500m)	10–10,000	0.01–10
Glacial activity		
Mountain	10–100	0.1–1
Greenland	100–100,000	0.1–10
East Antarctic	1000–100,000	10–100
West Antarctic	100–10,000	1–10
Water on land		
Groundwater	100–100,000	0.1–10
Lakes, Reservoirs	100–100,000	0.01–0.1
Tectonic changes		
Crustal formation, subduction	100,000–100,000,000	1–100
Glacial rebound	100–10,000	0.1–10
Continental collision	100,000–100,000,000	10–100
Sea floor, continental shifts	100,000–100,000,000	10–100
Sedimentation in ocean	10,000–100,000,000	1–100

other locations became submerged (a rise in sea level). A similar situation occurred in 1938 on the north island of New Zealand when an airport was almost completely submerged by the sea after an earthquake. There are also sea-level changes of lesser magnitude and over longer periods of time before and after earthquakes during historical time.

The movement of the crust can also produce tsunamis, i.e., seismic sea waves. Basically these phenomena are the result of changes that take place along the ocean basin such as displacement due to earthquakes, volcanic eruptions, or huge submarine slides. Events such as these are very rare in the Gulf of Mexico. They could be produced in the Caribbean along crustal plate boundaries with seismic sea waves moving into the Gulf of Mexico. These long and low waves move hundreds of kilometers per hour, speeds equivalent to those achieved by jet airplanes, across the ocean surface. As the wave enters shallow water its progression slows, it steepens,

Figure 1-1. Displacement of the earth's crust caused by the Alaskan earthquake of 1964; (left) change of more than 2 meters on a street in downtown Anchorage, (bottom) dropping of crust near the shoreline causing flooding of a house. Photographs courtesy NOAA.

and it can inundate the adjacent coast. This short-lived, local rise in sea level can be greater than that produced by a hurricane. The tsunami that took place in the northeastern Indian Ocean in 2004 was one of the most devastating in recorded history. Thousands of people were killed and it caused many billions of dollars of damage (Fig. 1-2).

Another crustal phenomenon that causes sea-level change is the formation or eruption of volcanoes. Like earthquakes, the addition of new crust associated with eruptions can cause a change in sea level. There is typically also some vertical movement of the crust on and adjacent to the volcano during an eruption. Although volcanoes were present in the Gulf of Mexico area during geologic time, there are no active volcanoes on or near the coast at the present time. There are, however, many areas of the Caribbean Sea where both earthquakes and volcanoes occur, such as the January 2010 earthquake in Haiti.

Lunar (astronomical) tides also provide for local, rapid, and significant changes in water level, although most people do not consider this important phenomenon as a sea-level change. The coasts of the world all experi-

Figure 1-2. Damage at Sumatra from the tsunami that occurred in the Indian Ocean in 2004 as the result of a major earthquake. Photograph courtesy Wikipedia.

ence lunar tides and do so at predictable and regular intervals. Lunar tides are caused by Newton's Law of Gravitation which states that bodies are attracted directly by the sum of their masses and inversely proportional to the square of the distance between them. In the case of the earth, the sun and moon are both major players in the phenomenon. Although the sun is many times the mass of the moon, the moon has about twice the influence on tides as the sun because it is so close. This attraction deforms the envelope of water around the earth, i.e. the oceans, thus forming the tides. The rotation of the earth causes the rise and fall of the tides and the precession of the moon results in spring or maximum tides and neap or minimal tides over a lunar month. Both spring and neap tides occur twice during a lunar month.

There is great range in the amount of water level change caused by astronomical tides. Many islands, such as the Bahamas, have tides of only a few decimeters in magnitude. On the other end of the spectrum is the Bay of Fundy, where water levels change 15 m during spring tides. All of the tides along the Gulf of Mexico coast are fairly low in magnitude (microtidal; <2 m) with the greatest being just over 1 m. These water level changes take place in periods of just over six hours (semi-diurnal tides) or nearly 13 hours (diurnal tides). Each estuary, embayment, or barrier island experiences different tidal conditions due to geographic differences in location and coastal morphology.

Sometimes tide gauge records show a change in sea level of several centimeters as much as a few months before and for several months after a major event. This is interpreted as slow crustal movement prior to or after an earthquake, the rapid and much larger movement during the earthquake, and the slow adjustment of the crustal position after the earthquake during which aftershocks occur. These types of rapid sea-level change due to crustal movement typically do not occur along the Gulf Coast. They are, however, very common along the Pacific coast of North America.

Weather also causes water level changes, typically called *wind tides.* This is the result of the friction between the wind and the water surface. It can take place in coastal bays or on the open coast. During the passage of hurricanes the term *storm surge* is typically applied to this phenomenon. This type of sea-level change is most commonly in the form of a rise in water level due to the friction of the onshore wind over the water surface. The opposite condition occurs when the wind blows offshore. The lowering of sea level is common along the northwestern Gulf of Mexico coast as cold fronts pass during the winter months. These frontal systems have strong winds from north to south just after passage of the front. The offshore blowing winds may cause a local lowering of sea level and may mask the small magnitude of lunar tides along the Gulf Coast (Fig. 1-3). The extent of these sea-level changes varies with the size and intensity of the storms. Very strong hurricanes can cause a regional scale change in water level of meters. Because lunar tides along most of the coast of the Gulf of Mexico are less than 1 m, it is commonly called a wind tide-dominated coast, especially in Texas.

Another atmospheric pressure phenomenon that causes water level change is a *seiche.* These are typically important in modestly sized water bodies such as the scale of the Great Lakes. Wind stress on the water surface causes an elevation of water level downwind and a depression of the water level on the upwind direction. A rapid decrease in wind speed will cause the water to essentially slosh back and forth across the basin; much like walking with a full coffee cup causes the surface to tilt back and forth. On Lake Michigan there have been deaths caused by unexpected and rapid increases in water level that swept people off the beach. Lake Erie has experienced water level differences of more than 3 m between the downwind and upwind ends of the lake. When the atmospheric pressure increases and the wind stops, the water surface in this relatively shallow lake will slosh back and forth like a large bathtub.

Withdrawal of fluids, such as water or hydrocarbons, from beneath the earth's surface causes a reduction in volume that produces a rise in sea level due to land surface subsidence. This is a major problem in some areas of the Gulf of Mexico and will be discussed in detail later in the book. Some areas of East Texas and Louisiana are especially vulnerable to this type of sea-level rise. Included are submerged agricultural areas (i.e., rice fields), residential areas, and roads (Fig. 1-4).

Compaction of muddy sediments can also cause a local sea-level rise. This is commonly associated with rapid accumulation of sediments such as in river deltas. These muddy sediments may contain up to 90% by volume of water that, when compacted, will be greatly reduced in volume and thereby cause sea level to rise in that area. Typically this compaction

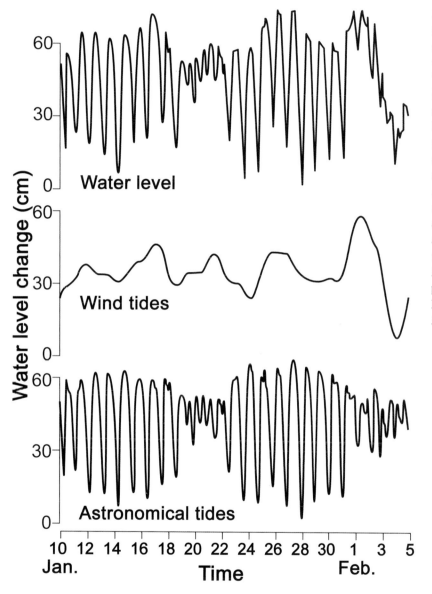

Figure 1-3. Offshore wind on the Texas coast has a significant influence on water level which can exceed lunar tides. This chart shows the water level changes on the top and the predicted tidal range on the bottom. The water level change in the middle is the difference of the two which is the result of wind; negative is offshore and positive is onshore. This record includes the passage of a major cold front and its associated increase in wind velocity and direction. From Davis, R. A., 1991, Oceanography, an Introduction to Marine Environments, 2nd ed., W. C. Brown, Dubuque, p. 118.

Figure 1-4. Submerged area around a house in the Galveston Bay area of Texas where fluid withdrawl has caused subsidence and subsequent sea level rise. Photograph courtesy O. H. Pilkey.

takes a long time, several centuries and more, but it might be both rapid and large in magnitude.

Regional Sea-Level Change

There are sea-level changes that take place over several hundred kilometers or more of a coast but not globally. The most prominent of these regional sea-level changes is that associated with widespread ice sheets and melting. In some areas of Antarctica the ice is more than 3 km thick generating a huge mass. If this ice were to melt, sea level would rise 60 m. This will not happen under present circumstances but even a small fraction of this ice melting can cause sea-level rise of a meter or two over the next few centuries. The huge mass of the glacial ice causes the crust to subside (isostacy) and therefore causes a relative rise in sea level. In addition, the ice mass causes a slight bulge in the crust in front of the moving sheet. Melting of these ice sheets results in reduction and eventual elimination of the weight, allowing the crust to rebound and approach its position prior to the advent of the ice sheet. These changes in the crust take place over millennia. The Scandinavian area of northern Europe is still rebounding from the melting of the most recent ice sheets. Sea level is actually falling along parts of the Norwegian coast (Fig. 1-5). As crustal rebound took place, sea level fell leaving a record of numerous shore-line positions on the land's surface (Fig. 1-6). Rebound has occurred in the northeastern United States, and especially in the Great Lakes region, but not in the Gulf of Mexico because the closest ice sheets were about 1000 km away.

Another type of regional sea-level change is associated with plate tectonics. As the plates move there is a vertical component to that movement as well as the more obvious horizontal motion. This takes place over millions of years (Table 1-1) and is masked by other more rapid changes. Examples of this would be along the west coast of North America and also along the north edge of the Caribbean Plate as evidenced by the earthquake in Haiti in January of 2010.

There are also annual cycles of sea-level change that are regional in scope. These may be as extensive as an entire ocean basin but are not global. The combination of oceanic circulation, astronomical relationships between the sun, moon, and earth and during some years, the El Niño phenomenon, all contribute to these changes. The El Niño cycles show changes in tidal range of up to about 20 cm (Fig. 1-7). Multi-year cycles can also be recognized by looking at long-term tide gauge records.

Figure 1-5. Diagram showing the crustal rebound in Norway after removal of the glacial ice sheets. As a result of continuing adjustment, sea level is still falling along much of this coast. Modified from Bloom, A., 1978, Geomorphology, Englewood Cliffs, NJ, Prentice-Hall, Inc. p. 405.

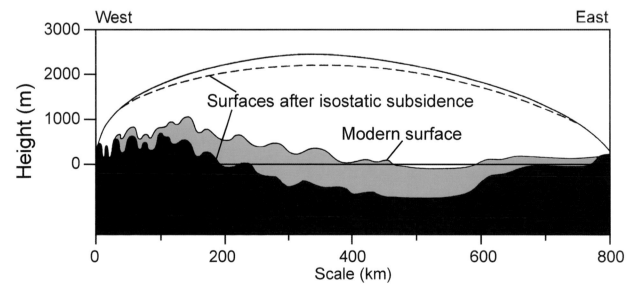

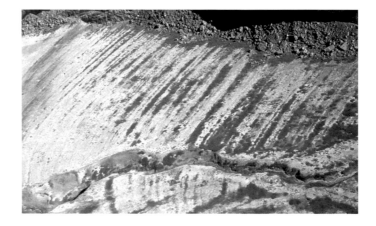

Figure 1-6. Aerial photograph of coastal area of Hudson Bay, Canada showing multiple beach ridges that developed as sea level was being lowered due to crustal rebound after melting of the glaciers. Photograph courtesy A. Hequette.

Figure 1-7. Changes in sea level caused by El Niño phenomena as recorded around the Pacific Ocean. Modified from Komar, P. D. and D. B. Enfield, 1987, Short-term sea-level changes and coastal erosion, In Sea-Level Fluctuation and Coastal Evolution, ed. Nummedal, D., O. H. Pilkey, and J. D. Howard, 17–28. SEPM Special Publication 41, Tulsa, Oklahoma Society of Economic Paleontologists and Mineralogists (now SEPM), The Society for Sedimentary Geology, p. 24.

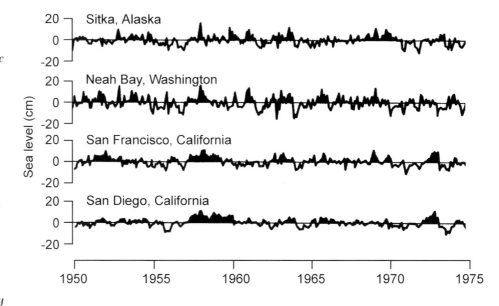

Eustatic Sea-Level Changes

Although the melting of ice sheets is the obvious and probably the most important cause of eustatic sea level at the present time, there are other factors that also make significant contributions. The basis for global sea-level change must lay in some combination of a change in the volume of water in the ocean and a change the volume of the ocean basins themselves. Ocean basins change as crustal plates move and oceanic ridges develop. This process is very slow, taking millions to tens of millions of years to result in enough difference in basin volume to produce significant sea-level changes. On the other hand, the formation and melting of extensive ice sheets takes place over hundreds to thousands of years (Table 1-1). A general sea-level curve over Phanerozoic time (from the Cambrian geologic period to present) shows that sea level has changed dramatically several times during the geologic past (Fig. 1-8).

An important mechanism of sea-level change is associated with plate tectonics. Movement of these plates is one of the most important factors in eustatic sea level change. As the plates move there is a vertical component to the movement of the plates as well as the more obvious horizontal movement. Such overall movement is on the order of a few centimeters per year. The end result is that the volume of the ocean basins is either reduced or an increased. For example, during the beginning of the Mesozoic Era about 180 million years ago, the crustal plates began separating. Over the next 10 million years or so the movement of the plates reduced the size of the ocean basin and produced a major period of sea-level rise that flooded numerous low elevation portions of continental masses. This process takes place very slowly (Table 1-1) and is masked by other more rapid changes.

The ocean surface is really not flat but has relief that is produced by currents, the earth's rotation, and persistent differences in atmospheric pressure across the globe. The centers of the oceanic circulation gyres tend to be areas of relatively high sea level and sea level is highest on the western side of the ocean due to the rotation of the earth. Atmospheric pressure is typically high in the center of these gyres thus depressing sea level (Fig. 1-9). The difference in water level from the high to the low pressure areas near the ocean basin margins is about 10 cm. In the Gulf of Mexico this difference is only 2–3 cm.

The ocean ridge system is the longest crustal feature on the earth. It extends for more than 50,000 km through all the oceans (Fig. 1-10). This feature is where new oceanic crust material is produced at spreading locations between the crustal plates. The spreading is the result of thermal circulation in the mantle. This ridge system is high in volcanic activity; Iceland straddles it at the present time. It is also an area where earthquakes are common. The oceanic ridge expands over many millions of years, in length, height, and width. In other words, it takes up a lot of

Figure 1-8. Sea level curve during Phanerozoic time (Cambrian to present) showing the many significant changes over hundreds of millions of years. Modified from Hallam, A.,1992, Phanerozoic Sea-Level Changes, New York, Columbia University Press, p. 4.

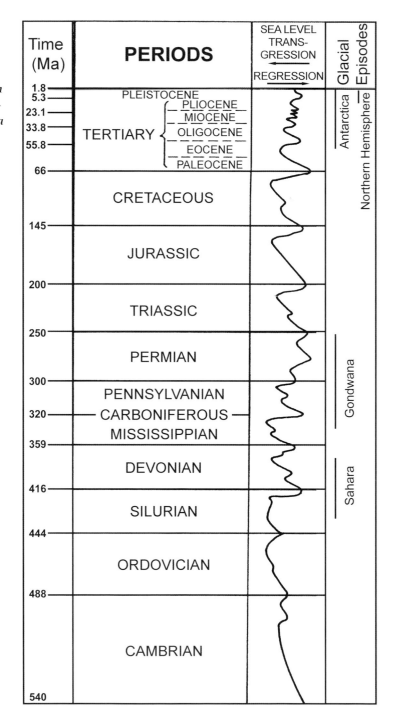

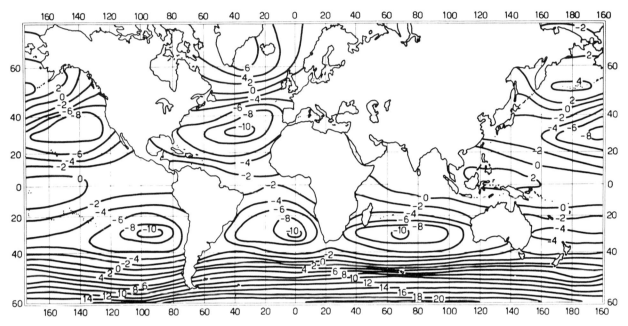

Figure 1-9. Global map showing the differences in sea level caused by atmospheric pressure (in cm). After Lisitzin, E., 1974, Sea-level Changes, Elsevier Science, Amsterdam, p. 61.

space in the ocean basins which results in a rise in sea level. The result is huge changes in ocean basin volume that produce the greatest changes in sea level of any phenomenon, up to several hundred meters, much more than fluctuations in ice sheets (Table 1-1). We are not typically aware of these changes in sea level because the rate of change is about 10^{-4} cm/yr and they take place hundreds of thousands of times slower than eustatic sea-level change caused by advances or retreats of ice sheets.

The sea-level change that has the world's attention at the present time is that related to global warming and melting of ice sheets and glaciers (Table 1-1). Currently, melting ice sheets and glaciers are causing sea level to rise, but if glacial ice was increasing, sea level would fall. The earth has been subjected to extensive but sporadic glacial activity for at least a billion years, but there is only good information on sea-level changes during the glacial episodes of the past 2.5 million years. Based on that abundant data, sea level has changed by as much as 130 m over the last 2.5 million years with numerous rises and falls. The maximum rate of change was about 2 cm/yr but it was more commonly only 1–2 mm/yr. This type of eustatic sea-level change will be discussed in detail later in the book.

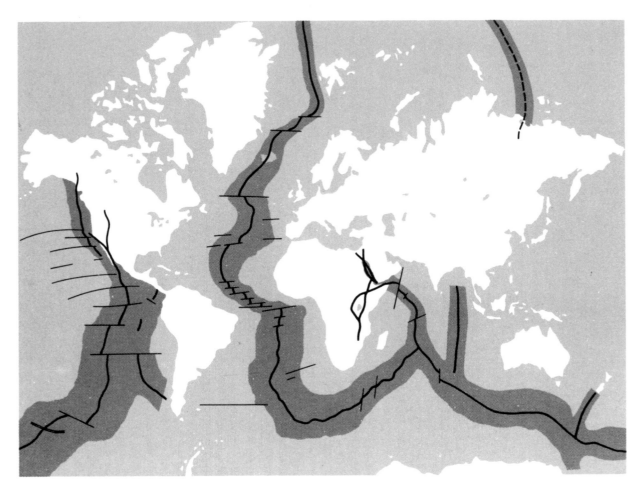

Figure 1-10. Map showing global distribution of the oceanic ridge system, the longest continuous feature on the earth's crust. The ridges mark the plate boundaries of the earth's crust.

Other minor changes in eustatic sea level that took place in the past are still making contributions to the present situation. In the 19th century, some people thought a major contributor to sea-level rise was the transport of sediments by rivers from land to the oceans, thus tending to fill the ocean basins. This has since been shown to be a very minor factor over the past few million years (Table 1-1), so small in fact that it is difficult to measure accurately. However, if the entire period of geologic time during which the Gulf of Mexico was in existence is considered, sedimentation in the basin is significant.

Beginning in the 20th century, thousands of dams were built to impound water for irrigation and for public water supplies (Fig. 1-11). These impoundments hold a significant amount of water that was headed to the ocean via rivers (Table 1-1). As a result of the capture of this huge total water volume, sea level was lowered. Extensive use of river water and reservoirs for irrigation deprives the oceans of water and therefore can contribute to a lowering of sea level. Globally, the amount is very small but it is significant. In contrast, a number of other anthropogenic activities have increased runoff from land and adding to the volume of ocean water. These include deforestation, paving roads and parking lots, groundwater mining, and water produced by combustion of fossil fuels. The capture of sediment behind rivers is also a factor in sea-level change because it reduces the amount of sediment that is carried to the ocean basins.

Another factor in eustatic sea-level change related to global warming is the increased volume of ocean water as it warms. Cooling of the water causes its volume to decrease. This is a contributor to the current rise in sea level (Table 1-1). The temperature change in the oceans takes place primarily in a thin upper layer of water that lies above the thermocline, a vertical distance of only a few hundred meters. Below that depth the water is at the temperature of maximum density, about 4° C. Because the

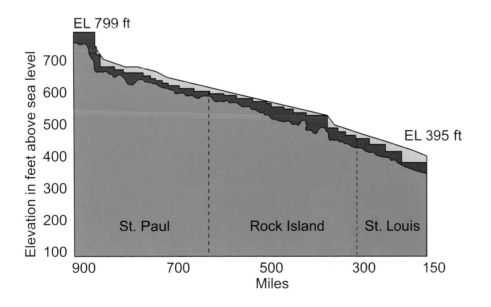

Figure 1-11. Locations of multiple dams along the upper Mississippi River system to its confluence with the Ohio River; an example of the widespread construction of such dams the result of which is a slight lowering of sea level due to the extensive impoundment of water and the capture of much sediment, some of which might reach the Mississippi Delta. Data from U.S. Army, Corps of Engineers.

average depth of the ocean is about 4,000 m, this thin layer of warming water has little influence on sea-level change.

Summary

It is obvious from the above discussion that there are many factors that contribute to sea-level change. Some are very local in nature and others take place throughout the globe. The rates and amounts of change range widely, from minutes to millions of years. Changes in sea level can be in excess of a hundred meters over thousands of years or only a small fraction of a millimeter over millions of years. Several of the local and short-term phenomena are spectacular. All of these factors should be considered when discussing sea-level change. It is easy to forget about some of them due to their slow and/or small influence in comparison to the melting of ice sheets that can have a major effect during only a few generations.

Important Readings

Douglas, B. C., M .S. Kearney, and S. P. Leatherman, (eds.). 2001. *Sea Level Rise: History and Consequences.* New York: Academic Press. Edited volume on sea-level problems with some important case histories.

Lisitzin, E. 1974. *Sea Level Changes.* Amsterdam: Elsevier. A broad but dated treatment of how and why sea level changed throughout geologic time by one of the leading experts.

Revelle, R., (ed.). 1990. *Sea Level Change.* Washington, D.C.: National Research Council, National Academy of Sciences. A nice overview of the processes, rates, and the record of past sea-level changes as was known at that time.

Titus, J. G. 1987. The causes and effects of sea level rise. In *Impact of Sea Level Rise on Society,* ed. H. G. Wind, 104–125. Rotterdam: A. A. Balkema. A good summary of the topic by the foremost federal government (NOAA) experts.

van der Plassche, O. (ed.). 1986. *Sea Level Research: A Manual for the Collection and Evaluation of Data.* IGCP Projects 60 & 200. Norwich, UK: Regency House. An edited volume that provides good information on how sea-level data are collected. It suffers a bit from being dated.

Sea-Level Changes in the Early History of the Gulf of Mexico

The Gulf of Mexico is relatively young in the history of the earth, yet it is still more than 150 million years old. Its origin is associated with plate tectonics as crustal plates moved thousands of kilometers. At times during its development, sea level in the Gulf of Mexico has been very high as well as low compared to the present. The history of this basin exhibits a lot of sea-level change. The Gulf of Mexico as it currently exists is only a few million years old even though the ancestral Gulf Coast is much older.

How can we tell that sea level changed in the distant geologic past? Actually it is pretty easy. In some cases there are inferences based on indirect data and in others, there are distinct records of these changes in the geologic strata and in the geomorphology of the past. Study of the data taken from thousands of cores across the states adjacent to the Gulf and on the continental shelf, and the seismic surveys across both land and sea have provided an excellent history of geologic development. The vast majority of these data have been collected by the petroleum industry.

Indicators of Ancient Sea-Level Change

The most obvious indicator of sea-level change during geologic time is the presence of marine fossils in sedimentary rocks at various locations and elevations throughout the continental land masses. Marine fossils can be found that are thousands of feet above present sea level. Most of the organisms that have been preserved as fossils lived in shallow marine environments on or near the continental shelves of the land masses. That places them within a few hundred feet below sea level at the time of their accumulation. We find these in Ordovician strata in Wisconsin (Fig. 2-1a,b) and many other locations that are now hundreds of meters above present sea level and hundreds of kilometers from the present

Figure 2-1. Fossiliferous limestone from the Ordovician of Iowa which is hundreds of meters above present sea level: (a) the strata and (b) a fossiliferous slab. Courtesy Carol Mankiewicz.

a

b

coastline. Ancient reefs are present in many locations in the rock strata such as in the Devonian of Michigan (Fig. 2-2a) and the Permian of West Texas (Fig. 2-2b) as well as in Appalachian and Rocky Mountain strata.

Another indicator for ancient sea-level change is the presence of sediments deposited by or in association with glaciers or ice sheets. These glaciogenic sediments are found in strata more than a billion years old. The presence of ice sheets in ancient times had to have resulted in significant changes in the volume of water in the ocean as ice formed the extensive ice sheets. This condition would, in turn, result in significant increases and decreases in sea level as ice masses formed or melted.

In undisturbed strata, abundant and widespread shallow marine fossils indicate sea level much higher than the present over extensive areas. Some fossiliferous strata have however, been moved vertically by mountain building (orogeny, uplift) and other tectonic processes. Numerous and widespread stream deposits suggest that sea level was low during the time when the stream-derived sediments were deposited. Streams cut into adjacent sediment layers as sea level falls, increasing downcutting by lowering the base level of the stream. This erosion produced abundant sediment that eventually was deposited in these deep river valleys and some of it made its way to the sea.

These types of evidence cannot provide any indication of how much sea level has changed, whether it was moving up or down at a specific time, or the rate of change, only that sea level was much different then than it is at its present position. Sea-level curves show the position of the ocean surface relative to the present over geologic time as compared to the present. These curves have been compiled using fossils, glaciogenic sediments, and other sea-level indicators such as unconformities (Fig. 2-3). An unconformity is a surface that represents a significant break in deposition and is characterized by erosion, generally subaerial erosion. It can be produced by changes in sea level and/or tectonics within crustal plates.

Much of the recent work on sea-level curves was done by a group of geologists at the Exxon research laboratory (now Exxon-Mobil) under the leadership of Dr. Peter Vail. They categorized sea-level changes as cycles of various durations. Examples of the longest or first-order cycles were the Paleozoic and the Mesozoic-Cenozoic eras that were each hundreds of million years in duration (see Fig. 1-8), and during which there were several rises and falls of eustatic sea level. There were 14 second-order

Figure 2-2. Paleozoic reefs that are now many hundreds of meters above sea level; (top) a Devonian reef of the Michigan Basin and (bottom) the Capitan Reef from the Permian in the Guadalupe Mountains of west Texas. Bottom photograph courtesy U.S. National Park Service.

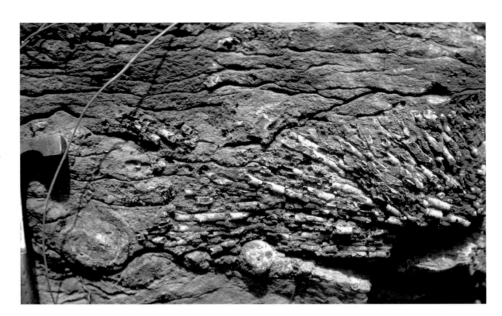

Figure 2-3. Photograph showing an unconformity in Cretaceous limestones of west Texas.

cycles that were tens of million years long. Each of these also included multiple incursions and excursions of sea level. There were more than 80 third-order cycles, each with durations of less than 10 million years. These cycles provided a hierarchy of sea-level change over Phanerozoic time, about 600 million years in total. Sea-level changes in the Gulf of Mexico have occurred only during Mesozoic-Cenozoic time, the last of the first-order cycles and a period of about 150 million years.

Geologists of the early 20th century used the percentage of land covered by rocks that represent a particular depositional environment to estimate the difference between present sea level and that of the time when these strata accumulated. For example, if half of the continents were covered by marine sediments, the amount of sea-level rise would be about 100 m because about half of the continent's surface is below that elevation. The next step in trying to recognize and quantify ancient sea-level change was derived from the investigation of cyclic strata that represented changes from coastal marine to stream deposits. These cyclic strata are a few to perhaps 100 m thick, thereby making it possible to estimate rates of sea-level change. These changes in sea level were mostly the result of advance and retreat of ice sheets.

Figure 2-4. Wave cut terrace on the coast near Sydney, Australia, an indicator of sea level position.

Figure 2-5. Sea stack in limestone with adjacent wave-cut platform near Havana, Cuba. These stacks are erosional remnants from erosion during Pleistocene sea level Stage 5e (125,000 years BP) which was about 3–5 meters above present sea level.

More than a century ago, scientists recognized beach deposits elevated much above present sea level. Other geomorphic indicators of sea level include wavecut terraces that form at or near the shorelines (Fig. 2-4), sea stacks (Fig. 2-5), buried coastal cliffs, sea-level notches (Fig. 2-6), and beachrock (Fig. 2-7). All of these features indicate that sea level was at

Figure 2-6. Notch in limestone showing position of sea level at the time of formation.

Figure 2-7. Beachrock on the Florida coast near Sarasota. This beachrock is a near equal mixture of quartz sand and shell debris and was lithified while in the beach environment.

or near these surfaces at the time they were formed. One of the problems associated with this line of research is that it is very difficult to accurately date the features. Beachrock can be dated easily because it is composed of calcium carbonate, however only materials less than about a million years old are datable using current methods. All of these types of evidence for sea-level change can be applied to the coastal region of the Gulf of Mexico.

Age and Origin of the Gulf of Mexico

Plate tectonics is the name given to the movement of continental crustal plates over denser oceanic crust. The movement of the plates is driven by thermal convection in the upper mantle, i.e., heat moves upward through the oceanic crust and forms convection currents that spread in opposite directions within the crust. This spreading location is the site of an elongate oceanic ridge system where volcanoes (e.g., Iceland) and earthquakes are common. At the beginning of the Mesozoic Era, the continental crust of the earth was together in one huge land mass, the supercontinent Pangea (Fig. 2-8). The strong thermal convection in the upper mantle caused the supercontinent to separate into several continental crustal masses. These crustal masses eventually became the continents that exist today. This episode of plate tectonics began in the Triassic about 180 million years ago. The crustal masses move only about a centimeter or so each year but over tens of millions of years this is many hundreds of kilometers. Such movements cause major changes in the size and shape of the ocean basins and therefore, major changes in sea level.

One of the most important features in the stratigraphy of the coastal plain of the Gulf of Mexico is the presence of several unconformities. These surfaces can tell us a lot about changes in sea level. They commonly have relief instead of being smooth or planar because they represent former land surfaces that were exposed to the atmosphere and therefore subject to erosion. Unconformable surfaces also tend to be at a very small angle to the overlying strata.

Recognition of unconformities enables us to say that sea level fell, exposing the strata under the present unconformity to subaerial erosion. A subsequent rise in sea level would then permit younger strata to bury this erosion surface. These buried erosion surfaces can be recognized in strata that

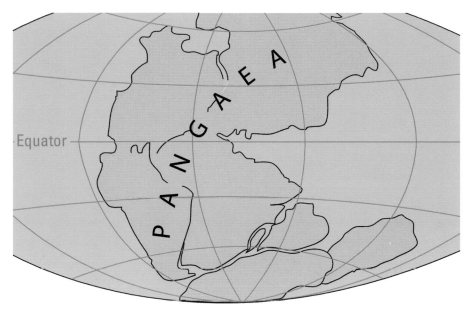

Figure 2-8. The super-continent Pangaea before the break up during the Triassic Period. Courtesy of U.S. Geological Survey.

PERMIAN
225 million years ago

crop out on roadcuts or along river valleys and also in the walls of quarries. Although more difficult to recognize, they can also be found in cores that are recovered from drilling through the strata. Currently, the most common approach to recognizing unconformities is through seismic surveys.

For the most part, the bulk of the stratigraphy of the coastal plain of the United States is rather simple with very gentle dips toward the Gulf (Fig. 2-9). There are places where normal faults exist and where changes from one type of lithology grades laterally into another type. The cross-sections shown in Figure 2-9 are simplified but they accurately convey the general trend of coastal plain stratigraphy.

Now that high-resolution seismic surveys are produced on both land and sea, these discontinuities are revealed in cross-sections. Seismic records are produced by the reflection of seismic waves from surfaces of strata. In these cross-sections the surfaces of the unconformities can be recognized (Fig. 2-10). Using seismic data, on land or sea, it is possible to estimate how much sea level changed from the time sediments began to be deposited over

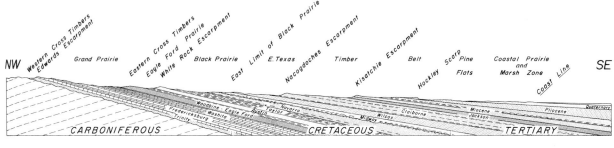

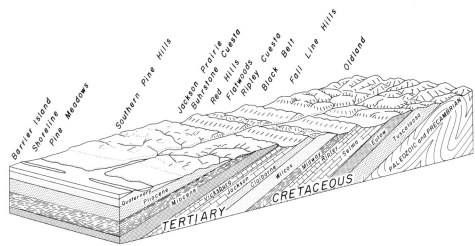

Figure 2-9. (top) Diagram showing the stratigraphy of the Texas coastal plain of the Gulf of Mexico and (bottom) a similar diagram showing the stratigraphy of the Mississippi/Alabama area of the coastal plain of the Gulf of Mexico. Both regions display a gentle Gulfward dip of the strata. Modified from G. Murray, 1961, The Atlantic and Gulf Coastal Province of North America. New York: Harper and Row.

the surface of the unconformity. The difference in elevation between each end of the unconformity provides some evidence for the minimum amount of sea-level change. The rate of change is much more difficult to determine from stratigraphic data such as seismic cross-sections. It requires determining the age difference between the seaward end of the unconformity and the landward end. Most of the time there is little or no datable material within the unconformity and the resolution of the age difference is insufficient for accurate determinations. In addition, tectonic activity can change the position and orientation of the unconformable surface.

The position of the unconformity relative to present sea level can also help determine the general position of sea level during the period of deposition of the strata above it. The sea-level curve for the Phanerozoic (see Fig. 1-8) shows several huge changes in sea level during the past 60

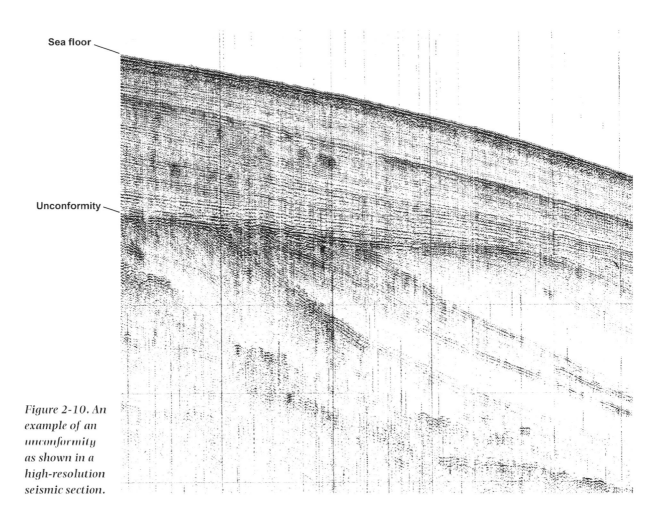

Sea floor

Unconformity

Figure 2-10. An example of an unconformity as shown in a high-resolution seismic section.

million years, the time during which the Gulf coastal plain was being deposited. Note that these curves only represent major fluctuations in sea level that took place over millions of years. There were also lesser changes that are not shown, largely because resolving their direction and magnitude is beyond our capabilities.

The formation of unconformities and the strata above and below them results from changes in sea level and the movement of the shoreline as sea level rises and falls. The landward movement of the shoreline that accompanies a rise in sea level is called *transgression* and the sediments that might accumulate during such a condition are packaged in a *transgres-*

sive sequence. The lowering of sea level and the seaward movement of the shoreline is *regression* and the associated strata are incorporated into a *regressive sequence.* The interpretation of these sequences and their respective depositional environments indicate the general scale of the sea-level change but not the precise amount up or down.

Early History of the Gulf of Mexico

Much of the data presented in this section are simplified and condensed from the work of Amos Salvador when he was at The University of Texas–

Figure 2-11. Initial conditions as what would become North and South America separated during the breakup of Pangea. This is about 170 million years before present during the Jurassic Period. Courtesy R. Blakey.

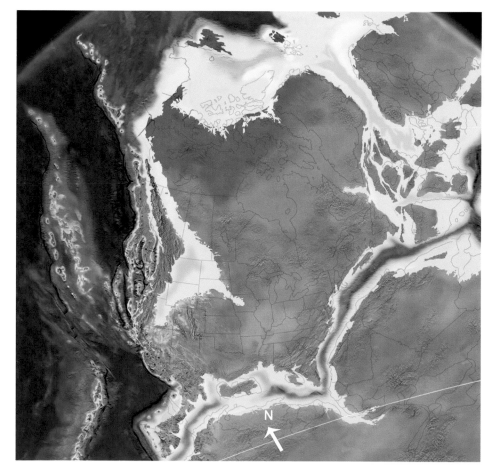

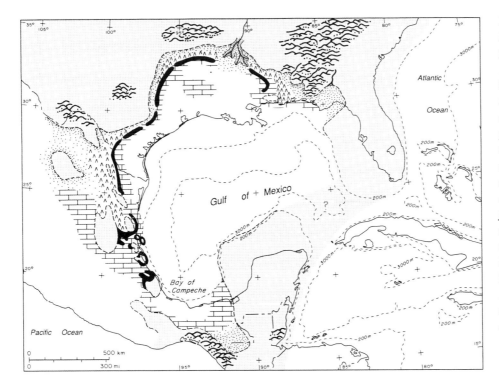

Figure 2-12. Gulf of Mexico region during Late Jurassic time about 130–140 million years before present. The initial appearance of the Gulf of Mexico took place here with deposition dominated by evaporate accumulation. Brick shapes are carbonate, dots are sand. Modified from Salvador, A., Origin and development of the Gulf of Mexico basin. In The Gulf of Mexico Basin, A. Salazar, ed., 389–444, Boulder, Colorado: Geological Society of America.

Austin. It was originally presented in the volume on the Gulf of Mexico published by the Geological Society of America as part of the series on *The Geology of North America* (see readings at end of chapter).

Mesozoic Era

The separation of the land masses that would became North and South America began about 170 million years ago during the Jurassic Period. As Pangea began to break up, a narrow belt of ocean was produced (Fig. 2-11). This area, now the Gulf of Mexico Basin, was a shallow, hypersaline sea where extensive salt deposition took place. The thick salt strata eventually produced many salt domes along the coast in the western Gulf of Mexico that are associated with numerous important oil fields.

The first time that tectonic stability prevailed in the Gulf was during the Late Jurassic about 130–140 million years ago (Fig. 2-12). At this time what are now the Yucatan Peninsula and the Florida Peninsula were isolated shallow marine masses in the form of carbonate platforms and the ancestral Gulf of Mexico was a shallow marine sea. The coast in Mexico and Texas was inland of the present coast and was dominated by reefs with shallow basins that precipitated evaporite minerals on their landward side. The ancestral Appalachian Mountains extended down to near the site of the present Florida Panhandle. These conditions put sea level hundreds of meters above that of the present time. This also began an extended period of crustal stability along the northern Gulf Coast although motion in the Caribbean Plate continued in the southeastern part of the basin. There was substantial subsidence of the basin during this time that eventually opened the Gulf between the Yucatan and Florida peninsulas. It is estimated that the rise of sea level during the Jurassic Period was about 150 m.

Carbonate deposition continued to dominate in the early Gulf of Mexico during the early Cretaceous Period and included large reef complexes throughout (Fig. 2-13). Landward of the carbonates in the northeastern

Figure 2-13. Gulf of Mexico region during Cretaceous time about 75 million years before present. This was a time of very high sea level and widespread carbonate deposition. Heavy black lines are reefs. Modified from Salvador, A., Origin and development of the Gulf of Mexico basin. In The Gulf of Mexico Basin, *A. Salazar, ed., 389–444, Boulder, Colorado: Geological Society of America.*

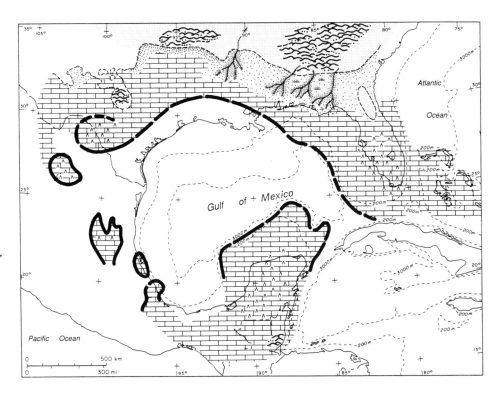

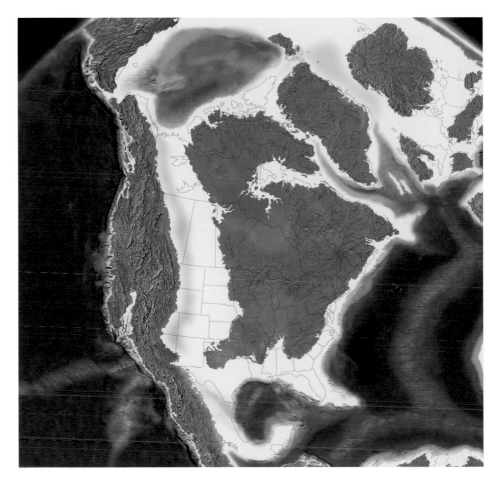

Figure 2-14. Reconstruction of North America and the Gulf of Mexico about 75 million years before present showing the Western Interior Sea that transects the entire continent and is connected with the Gulf. Courtesy R. Blakey.

Gulf of Mexico were the beginnings of land-derived (terrigenous) sediments coming from the southern Appalachians. The first deltas of the Gulf of Mexico Basin were deposited at that time.

The coastal plain of the Gulf of Mexico began to form during the latter portion of Cretaceous period about 75 million years ago. Considerable amounts of sediment were coming into the northeastern part of the Gulf from the southern Appalachians whereas the northwestern portion of the Gulf was accumulating carbonate sediments (limestone) in the absence of significant runoff from rivers. In fact, the Western Interior Seaway extended across North America to the Gulf of Mexico (Fig. 2-14). Carbonates also dominated the Yucatan and Florida platforms. Near the end of the Cretaceous Period, sea level fell and exposed many of the shallow margins and platforms where carbonate sediments had been accumulating. At that time there was substantial tectonic activity in other parts of the Americas associated with the formation of the Rocky Mountains and the Andes (i.e., the Laramide Orogeny). There was also some crustal activity in the Mississippi Embayment, the geologic region to the north of the Gulf of Mexico. As a consequence, the ocean basins experienced a significant increase in volume that produced lower sea level in the Gulf.

This lowered sea level caused significant erosion of adjacent land masses resulting in the delivery of substantial sediment throughout the northern Gulf of Mexico coast. Areas of volcanic activity included the Texas coastal area of that time and what is now northern Louisiana and northern Mississippi. While this was occurring, sea level was rising to provide shallow marine conditions and related carbonate sediment accumulation throughout the margins of the Gulf of Mexico Basin.

Several transgressions and regressions of the sea took place during the Cretaceous Period. Generally each of the transgressions was about two million years long followed by regressions of about the same length. Between each of these cycles was a time of widespread sediment accumulation; mostly carbonates in the north and west of the Gulf and land-derived (terrigenous) sand from the Appalachians in the northeast. During the late Cretaceous, the Florida Platform was accumulating carbonates and was separated from North America so that Appalachian sediment runoff was not transported to the platform. Our best estimate of the highest sea-level stand for the Gulf of Mexico, about 250 m above the present position, was near the end of the Cretaceous (Fig. 2-14).

Chuxilub Tsunami

The end of the Cretaceous Period was marked by a mass extinction of several large groups of organisms including the dinosaurs. For many years there was considerable debate about the cause(s) of these extinctions. It is now generally agreed that a huge impact event from an extraterrestrial body occurred on the Yucatan Peninsula (Fig. 2-15) and that this event caused changes to the earth that resulted in mass extinction. The site of this impact structure encompasses more than 100 km in diameter and is called Chuxilub. One of the important phenomena that re-

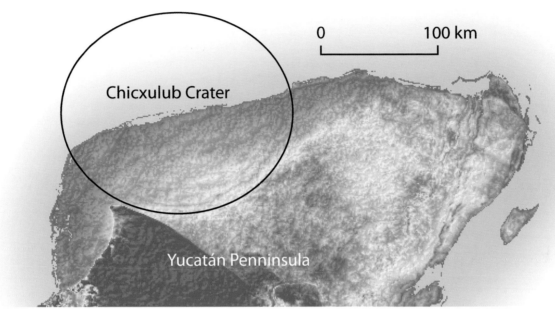

Figure 2-15. Yucatan peninsula as it is now configured showing the Chuxilub impact structure. Arcuate rings indicate topography formed by the impact. Figure by Anthony Reisinger with data from NASA, Shuttle Radar Topography Missison.

sulted from this impact was the generation of tsunamis that temporarily raised sea level around the Gulf of Mexico. Deposits from the tsunami, called *tsunamites*, have been recognized in various places around the Gulf and even in other parts of the world. Detailed studies of these have been made by numerous researchers including Professor David King of Auburn University and his colleagues. They found tsunamites nearly a meter thick in Alabama lying immediately above a bed with many tiny pieces of meteorite glass spherules. This would put the elevation of the tsunami waves about 100 m higher than the present sea level, an elevation never reached since that time.

Tertiary Period

The ocean basins, including the Gulf of Mexico, contain an excellent and extensive accumulation of sediments that were deposited during the Tertiary Period that began about 65 million years ago and lasted more than 60 million years. The older oceanic strata are more local in distribution. Analysis of these older strata shows that sea level made multiple incursions and excursions during the Tertiary Period (see Fig. 1-8). Overall, sea-level changes during the Paleocene and Eocene epochs of the Tertiary were small and short in duration. However, during the middle of the Oligocene Epoch there was an overall lowering of sea level.

During the Paleocene Epoch (see Fig. 1-8) researchers have identified sequences in the eastern Gulf coastal plain using bottom dwelling (benthic) Foraminifera which are single-celled, amoeboid protists. These sequences lasted less than a half-million years. The cause(s) of these cyclic sea-level changes is (are) problematical because there is no evidence of extensive glaciation during this time.

Little change in sea level is recorded in the rock record for the Eocene Epoch (see Fig. 1-8) except for an abrupt fall and rise at the beginning of the middle portion of the epoch (Fig. 2-16). The aforementioned major drop in sea level at the beginning of the mid-Oligocene lasted for mil-

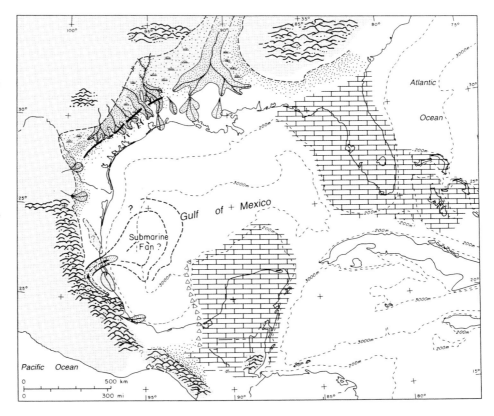

Figure 2-16. Gulf of Mexico region during Eocene time about 55 million years before present. Abundant deltas existed on the northern and western Gulf Coast. Modified from Salvador, A., Origin and development of the Gulf of Mexico basin. In The Gulf of Mexico Basin, *A. Salazar, ed., 389–444, Boulder, Colorado: Geological Society of America.*

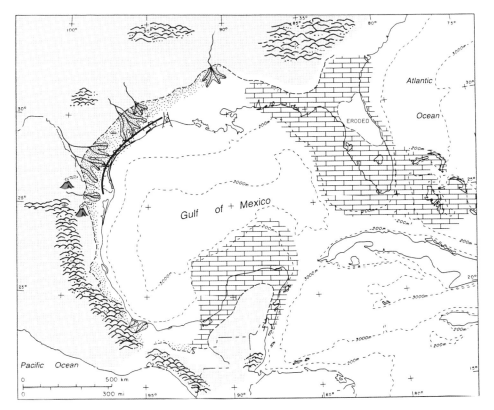

Figure 2-17. Gulf of Mexico region during Oligocene time about 35 million years before present showing continued presence of deltas. The climate was warm and the ocean basins were increasing in volume thus producing the decrease in sea level. Modified from Salvador, A., Origin and development of the Gulf of Mexico basin. In The Gulf of Mexico Basin, A. Salazar, ed., 389–444, Boulder, Colorado: Geological Society of America.

lions of years and caused a great deal of land-based erosion and sediment delivery to the northwestern Gulf Coast. During this time, the U.S. Gulf Coast was divided into two distinctly different provinces: a terrigenous-dominated complex of deltaic sedimentation to the west, and the extensive Florida carbonate platform on the east. The Mexican coast was primarily terrigenous sediments coming from the nearby mountains. Deltaic sediment accumulation was not significant on the Mexican coast because of the lack of major stream networks feeding the coast. The Yucatan carbonate platform dominated the southern coast of the Gulf (Fig. 2-17).

During the Miocene Epoch (see Fig. 1-8) there was extensive uplift of the Appalachian area which produced a considerable amount of sediment that was transported to the south. This sediment filled the Straits of Florida which had separated the Florida platform from the U.S. mainland

and was carried across what is now the Florida Peninsula. Large deltaic systems developed along the northern and northwestern coasts of the Gulf of Mexico basin. The Yucatan and Florida peninsulas continued to be dominated by carbonate sedimentation. However, sediment from the Appalachians was starting to encroach onto Florida as the straits between the platform and the continent closed to form a peninsula. Sea level was tens of meters above its present position (Fig. 2-18).

As the Miocene gave way to the Pliocene Epoch (see Fig. 1-8) this pattern continued with the U.S. Gulf Coast becoming dominated by siliciclastic (terrigenous) sediments from the mainland. Only the Yucatan platform continued to be dominated by carbonate sediment accumulation due to a lack of siliclastic sediment input (Fig. 2-19). The shoreline began to take on a configuration similar to the present time but at a more landward position.

Figure 2-18. Gulf of Mexico region during Miocene time about 20 million years before present. During this time, the straits between Georgia and Florida were filled with sediment from the southern Appalachians. This led to the first terrigenous sediments on the Florida peninsula. Modified from Salvador, A., Origin and development of the Gulf of Mexico basin. In The Gulf of Mexico Basin, *A. Salazar, ed., 389–444, Boulder, Colorado: Geological Society of America.*

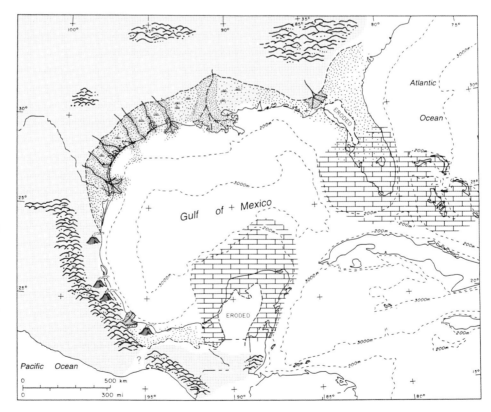

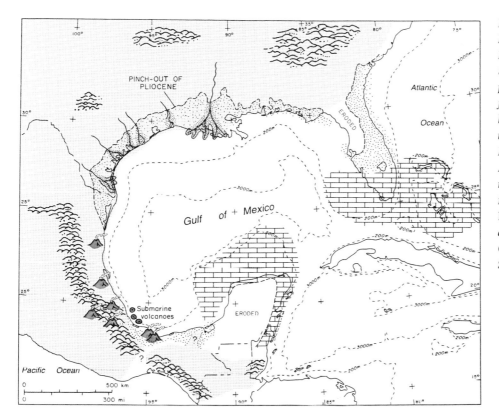

Figure 2-19. Gulf of Mexico region during Pliocene time about 8 million years before present. Ice sheets were forming in the southern hemisphere leading to the lowering of sea level. Modified from Salvador, A., Origin and development of the Gulf of Mexico basin. In The Gulf of Mexico Basin, A. Salazar, ed., 389–444, Boulder, Colorado: Geological Society of America.

Summary

It has taken more than 100 million years for the Gulf of Mexico Basin to develop, but its development has been geologically rather simple and constant. Sea level made several incursions and excursions ranging from the highest in the Cretaceous Period where carbonate accumulations extended to elevations of more than 100 m above present sea level in places like central Texas in the northwest and onto the Florida and Yucatan platforms to the lowest only tens of thousands of years ago as Pleistocene ice sheets caused a drop of more than 100 meters below the present sea-level position. The changes in the Pleistocene will be discussed in detail in the next chapter. In between these highest and lowest stands, large river systems carried huge amounts of sediment from land to the coast where they accumulated as large delta systems. This was accompanied by a sea-level fall and an advancement of the extensive coastal plain. Within this big picture of sea-level change there were many ups and downs of sea level with periodic times of stable sea-level conditions.

Important Readings

Hallam, A. 1992. *Phanerozoic Sea-Level Changes.* New York: Columbia University Press. A comprehensive look at sea-level change through a large portion of geologic time.

Haq, B. U., J. Hardenbol, and P. R. Vail. 1987. Chronology of fluctuating sea levels since the Triassic. *Science* 235:1156–167. A very excellent paper that provides details about sea-level fluctuations during the portion of geologic time during which the Gulf of Mexico was present.

Murray, G. E. 1961. *The Atlantic and Gulf Coastal Province of North America.* New York: Harper and Row. The first really comprehensive regional stratigraphy volume on the Gulf of Mexico coast.

Salvador, A. 1991. Origin and development of the Gulf of Mexico basin. In *The Gulf of Mexico Basin,* A. Salvador, ed., 389–444. Boulder, Colo.: Geological Society America. A really complete treatment of the geologic development of the Gulf of Mexico over about the past 180 million years. Much of this chapter is based on the work of Salvador.

Vail, P. R., R. M. Mitchum, Jr., and S. Thompson, III. 1977. Seismic. stratigraphy and global sea level changes of sea level part 4: global cycles of relative changes in sea level. In *Seismic Stratigraphy—Applications to Hydrocarbon Exploration,* ed. C. E. Payton, 83–97. AAPG Memoir 26. Tulsa, Okla: American Association of Petrology Geology. The first volume of research papers dealing with the concept of sequence stratigraphy. This approach has completely transformed how we interpret the history of continental margins.

Wilgus, C. H., B. K. Hastings, B. Posamentier, J. Van Wagoner, C. A. Ross, and C. G. St.C. Kendall, (eds.). 1994. *Sea-level Changes—An Integrated Approach.* SEPM Special Publication No. 42. Tulsa, Okla: Society for Sedimentary Geology. This is one of the classic collections of papers on sequence stratigraphy, nearly all of which comes from the Gulf of Mexico. In a way, it is a more sophisticated sequel to the Vail et al. volume.

3

Sea-Level Changes During Glacial Times

Glaciers are large masses of moving ice. They can be small such as those in the mountain valleys in the Alps and Alaska or they can be very widespread such as those covering most of Greenland and Antarctica. These very large glaciers are know as *ice sheets*, It is these ice sheets that have the potential to change sea level as they grow and melt. The East Antarctic Ice Sheet formed by the end of the Eocene Epoch (see Fig. 1-8) and the West Antarctic Ice Sheet began to form in the middle Miocene. As a result, these ice sheets regulated glacial eustacy prior to the development of the ice sheets in the Northern Hemisphere. Beginning about 2.5 million years ago, ice sheets became prominent in the Northern Hemisphere and subsequently became widespread over much of the earth's surface in response to cooling global climate. The Quaternary Period of earth history began about 2.6 million years ago. The period is divided into the Pleistocene Epoch and the Holocene which began about 10,000 years ago. In this discussion the focus is the late Pleistocene that began about 800,000 years ago.

These ice sheets of the late Pleistocene covered millions of square kilometers and many were up to 3 km thick. The water that produced all of this ice originated in the oceans and was carried to the continents as snow. Although ice sheets did not come within a thousand kilometers of the Gulf of Mexico, the huge volume of water that was removed from the oceans produced a major fall in global (eustatic) sea level. The result was that sea level in the Gulf was about 130 m below its present position during periods of maximum glaciation. This caused most of what is now the Gulf of Mexico continental shelf to be exposed as part of the coastal plain. The Florida and Yucatan carbonate platforms were also exposed and developed karstic surfaces with numerous sinkholes. Sea level changed from the Cretaceous Period about 100 million years ago when it was the highest in the history of the Gulf of Mexico to the most recent glacial advance of the late Pleistocene, only about 20,000 years ago, when it was its lowest.

Pleistocene Glacial History

It takes only a change of a few degrees Celsius in average global temperature to cause extensive development of ice sheets. Scientists believe that the global average temperature of only 2–3° C less than the present caused the extensive ice sheets of the Pleistocene. Such changes have taken place multiple times over the history of the earth but were only significant during one period in the history of the Gulf of Mexico. What happens in the earth system that causes such changes in climate? It must be associated with our relationship to the sun.

In 1924 a Serbian astronomer by the name of Milutin Milankovitch worked out the cycles of the changes in the earth's path and orientation relative to the sun that causes changes in climate. He determined that the shape of the earth's orbit changes from essentially circular to elliptical over a period of 96,000 years. As a result there is a long-term variation in the amount of heat the earth receives from the sun as the distance between the two changes cyclically. He also found that the tilt of the earth on its axis ranges from 22.1° to 24.5°. As we learned in middle-school geography, it is now 23.5°, which is the current position of the Tropic of Cancer and the Tropic of Capricorn. This change in latitude takes place over a 41,000 year cycle and the greater the tilt, the greater the amount of sunlight and heat that is received at the poles. This causes the seasons to be more extreme, similar to what we are now experiencing with colder winters and hotter summers. The last of the three cycles is concerned with the wobble of the earth on its axis as it moves in space. This cycle is 21,000 years long and determines during which time of year the earth is closest to the sun (Fig. 3-1). We now accept the Milankovitch cycles as controlling factors in climate change that have been in effect over the history of the earth. The superimposition of these cycles leads to complicated phase relationships that determine climate conditions (Fig. 3-2).

The ice sheets that resulted from the climatic effects of these cycles during the Pleistocene Epoch extended across Canada and much of the northern United States in North America and most of northern Europe.

Milankovitch Cycles

A - Eccentricity

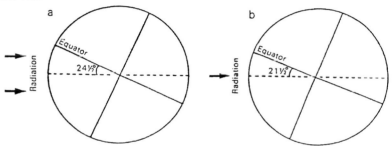

B - Axial Tilt

C - Precession of the Equinoxes

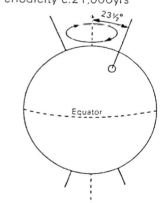

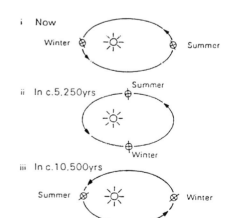

Figure 3-1. Sketches of the three Milankovitch cycles with the astronomical relationships showing the length of each.

Figure 3-2. Timeline for the superimposed cycles of Milankovitch showing their relationships to glacial episodes over 200,000 years. Stadial periods are shorter than glacial periods and are relatively cold whereas interstadial periods are warm.

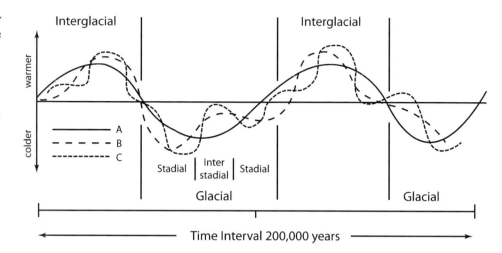

There were smaller ice sheets and ice caps in Asia and South America. All of Greenland and the Antarctic continent were also covered with ice. Early workers studying land-based deposits and geomorphology recognized that the ice sheets advanced and retreated at least four times in North America in response to the Milankovitch cycles. The first two glacial stages, called the Nebraskan and Kansan, were most extensive in these plains states but also covered much of the other more eastern states. The latter two, the Illinoisan and Wisconsinan, extended further south and were more extensive in their coverage. The southernmost extent of the Wisconsinan glaciation went as far south as near the Missouri and Ohio rivers including the confluence of the Ohio and Mississippi rivers at the southern tip of Illinois. Four stages of ice sheet advance and retreat are also recognized in Europe but are named differently (Table 3-1).

Advances of the ice sheets caused huge falls of eustatic sea level. Between each glacial period was an interglacial stage when melting occurred (Table 3-1) causing a large rise in sea level. These major cyclic changes in global sea level dominated the Quaternary Period of geologic time and have been prominent in the development of the present coastal region of the Gulf of Mexico.

While a doctoral student at the University of Chicago in the early 1950s, Cesare Emiliani developed an approach to the paleoclimatic investigation of oceanic sediments that has provided a much more detailed account of the glacial and climatic history of the Quaternary Period. Deep

TABLE **3-1. Stages of Pleistocene glaciation.**

North America		Alps	
Glacial	*Interglacial*	*Glacial*	*Interglacial*
Wisconsinan		Würm	
	Sangamon		M/W
Illinoisan		Mindel	
	Yarmouth		R/M
Kansan		Riss	
	Aftonian		G/R
Nebraskan		Gunz	

sea sediments accumulate continuously as fine-grained terrigenous sediments and skeletal material of microscopic organisms settle to the bottom of the ocean. These sediment accumulations in the deep ocean have not been eroded therefore they provide an undisturbed record of oceanic history. Analysis of these sediment sequences provides detailed information on sea-level changes during the Quaternary. The technique involves determining a proxy for oceanic temperature and thus for global climatic conditions. These data also allow the volume of ice on the planet to be estimated. This proxy is determined by analyzing details of the chemical composition of the skeletal material of the microscopic organisms which are composed of calcium carbonate ($CaCO_3$).

Oxygen exists in two different forms or isotopes, oxygen-16 (O^{16}), the most common, and oxygen-18 (O^{18}). These isotopes are present in different ratios in ice as compared to normal sea water. When the earth's climate becomes cold, seawater becomes enriched in O^{18}. The microscopic organisms analyzed using Emiliani's techniques are planktonic (floating) Foraminifera that live in the upper waters of the open ocean. The tests (shells) of these single-celled organisms are composed of calcium carbonate. The oxygen that is incorporated into these carbonate tests is present in the same ratio of O^{16} to O^{18} that occurs in seawater. The colder water temperature changes the ratio of the oxygen isotopes in the tests of the organisms. When there is considerable ice on the ocean surface, the oxygen in the meltwater from this ice contains more O^{18}. The Foraminifera incorporate this O^{18}-enriched water into their tests. When the organisms die, their tests settle to the ocean floor and are incorporated into the sediments that accumulate over time. Therefore, the tests of Foraminifera can be extracted from sediment cores, the oxygen isotope ratios of the tests can be measured, and paleoclimate characteristics for the earth can be determined.

The paleoclimate data collected allow changes in sea level that result from climate changes as measured by oxygen isotope ratios to be determined. Oxygen-isotope analyses have been conducted throughout the world's oceans and have resulted in a detailed record of sea-level change (Fig. 3-3). These isotope stages, as they are called, are identified by number with the

Figure 3-3. (a) Graph showing the sea level positions based on oxygen isotope data for the past 2.5 million years and (b) the expanded most recent part of the Pleistocene sea level record. Modified from Shakleton, N.J., A. Berger, and W. R. Peltier, 1990, An alternative astronomical calibration of the lower Pleistocene timescale based upon ODP site 677. Transactions of the Royal Society of Edinburgh, Earth Sciences. 81:251–61.

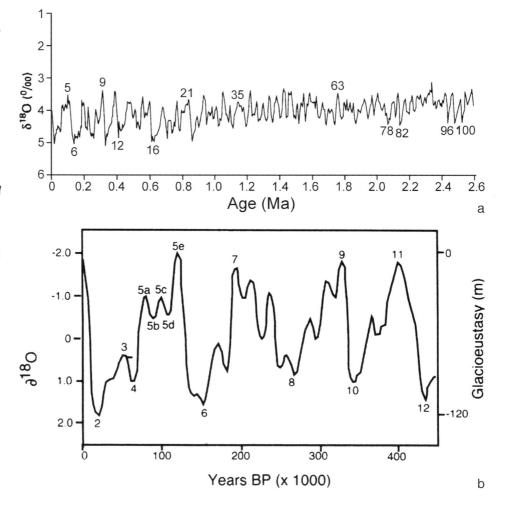

lowest numbers representing the youngest stages. Odd numbers indicate sea-level highstands that correspond to glacial retreats and the even numbers represent sea-level lowstands that correspond to glacial advances. In addition to the primary cycles or stages, there are shorter term and smaller changes called substages. These are labeled by letter. For example Stage 5 is divided into multiple substages with the highest sea level of that stage represented by 5e (see Fig. 3-3). Substage 5e will be discussed more later.

Ups and Downs of Sea Level

The techniques pioneered by Emiliani were later refined by Nicholas Shackleton of Cambridge University in England, and have been the most important development in scientists' ability to chronicle the story of sea level during the Pleistocene glaciations. These techniques allow Pleistocene changes in eustatic sea level position to be determined over as little as a few thousand years. This contrasts with the relative lack of detail on sea-level changes that occurred during earlier periods of geologic time. That is to be expected because of the relatively undisturbed sediment sequences, the well-preserved material in the ocean, and the geologically young age of the sediments.

A look at the chronograph for the glacial period (see Fig. 3-3) shows the details of sea-level change intervals as determined by the oxygen isotopes. For example, the most recent high sea-level stage (5) occurred between about 80,000 and 130,000 years ago (see Fig. 3-3b). This generally corresponds with the Sangamon interglacial period of our land-based chronology (Table 3-1). Note that there is an Early Wisconsinan and a Late Wisconsinan period of lowered sea level. The highstand labeled 5e is within the Early Wisconsinan, 125,000 years ago. This example demonstrates the more specific age determination from the sediment taken from the ocean floor as compared to land-based age determinations. As a consequence, land-based terms are not used when dealing with specific sea-level changes or positions.

Isotope Stage 5e is the most recent time during which sea level was significantly above its present position. It was about 3–5 m above the present level and produced old shorelines that are now preserved throughout many of the coastal areas of the Gulf of Mexico. These Stage 5e shore-

Figure 3-4. Stage 5e coral reefs just above present sea level on the coasts of (a) Cuba, (b) the Yucatan Peninsula of Mexico (photograph by G. Haralson), and (opposite page) (c) close up of Stage 5e corals.

a

b

c

lines are typically old beach/dune/barrier complexes that were produced under wave-dominated conditions, much like the present but typically without an open-water, backbarrier (estuarine) environment between the shoreline and the mainland.

In addition, there are several places on the south and southwestern Gulf Coast where Stage 5e coral reefs have been preserved and now occupy a position slightly above present sea level. These conditions are associated with the Florida Keys, the north coast of Cuba (Fig. 3-4a) and the Yucatan Peninsula (Fig. 3-4b, c). In the Yucatan there are several places where Stage 5e reefs are exposed including the mainland and the islands of Isla Mujeres and Cozumel. Although we assign a sea level of 3–5 m above the present level, the reef surfaces are less than that because the corals were submerged when they were alive. The reef surfaces have also been exposed for nearly 100,000 years so there has been some solution on the limestone surface.

In the Gulf of Mexico the formation, advance, and retreat of ice sheets produced important sea-level changes. As a consequence, these changes affected the Gulf of Mexico coast. There were undoubtedly local and regional changes that also occurred, especially in the area of the ancestral Mississippi Delta, but these are very difficult to extract from the more extensive and large basin-wide changes.

The coastal environments of the present Gulf of Mexico owe their origin to the sea-level changes that occurred during the late Pleistocene. As far as we know, in the history of the earth these sea-level changes were the most dramatic over a short period of geologic time. The advance and retreat of sea level over the Gulf Coast transported sediment, changed environments, and modified the climate of the region. Rates of sea-level change, up to a centimeter per year and a meter per century, were several times those that have been experienced during human occupation of the coast, the last 10,000 years or so.

As the first ice sheets expanded over the landmasses, sea level fell by many tens of meters causing the shoreline to move tens of kilometers seaward. As the result of this change, the landmass adjacent to the Gulf was increased by about 100,000 km². The consequences of this change were extensive. The base level of rivers that drained the coastal plain was lowered causing increased erosion, deepening their valleys and extending their drainage across the now wider coastal plain. These rivers carried huge quantities of sediment (Fig. 3-5).

Data from throughout the Gulf of Mexico and the adjacent coastal plain show that the present sea level is much closer to its highest position during the Pleistocene than to its lowest position. This is to be expected because of current ice sheet melting. The highest that sea level has been during the past million years or so is about 30–40 m above the present, as shown by old shoreline and coastal dunes in central Florida. Remember, however, that the outer coastal plain is very flat so that such an increase in sea level would drown all the coastal cities of the Gulf. By contrast, the lowest position of sea level during the same time period was about 130 m below its present position, moving the shoreline up to 150 km seaward of its present position depending on location. The position of that shoreline would be near the outer edge of the continental shelf. During that time of low sea level, rivers extended across the present shelf and much of the sediment they carried was transported to the edge of the shelf and cascaded down the slope in the form of suspended sediment called *turbidity currents* and other types of sediment gravity flow. As sea level rose after the ice sheets began to melt, the gradient of the rivers decreased and river valleys

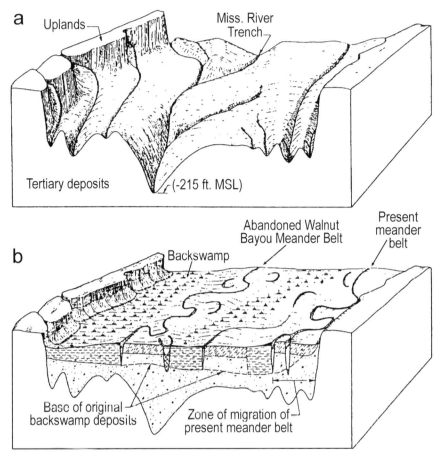

a Uplands   Miss. River Trench

Tertiary deposits   (-215 ft. MSL)

b Backswamp   Abandoned Walnut Bayou Meander Belt   Present meander belt

Base of original backswamp deposits   Zone of migration of present meander belt

Figure 3-5. Diagrams showing the Mississippi River valley (a) during a sea level lowstand when glaciers were advancing and (b) after sea level rose when sediments were again accumulating into this valley. From: Fisk, H. N., 1944, Geological Investigation of the Alluvial Valley of the Lower Mississippi River, Vicksburg, U.S. Army, Corps of Engineers, Mississippi River Commission.

were filled with sediment causing more lateral erosion and less sediment to reach the mouth of the rivers.

Products on the Coast Caused by Glaciation

Melting of the ice sheets caused existing rivers to enlarge and new rivers to form. These rivers transported huge amounts of sediment, sand and mud, to the coast. The rise in sea level associated with this melting provided a great deal of space for sediment to accumulate on the continental shelf. This is called *accommodation space.* The combination of increased runoff with accelerated sediment production resulted in a decrease in space for sediment to accumulate. As ice sheets expanded again and sea level fell, space on the shelf was reduced but more space was provided beyond the shelf where sediment could accumulate. There was obviously a significant lag between these cycles so that for a rather long time period, there was considerable space for the coast to prograde (build seaward) and sediment was transported to and accumulated in various environments.

River, or fluvial, environments dominate coastal plain sediment deposits. The shoreline was irregular as sea level was falling producing embayments where the rivers emptied and deposited deltas at their mouths. The embayments were brackish (low salinity) estuaries. In the other phase of the cycle, sea level advanced causing a decrease in river gradient. The flat coastal plain surface allowed the streams to meander and accumulate considerable sand. Flood plains were also present where mud was dominant, but as the meanders wandered across the coastal plain the mud was eroded and sand was deposited in its place. The result of these conditions

Figure 3-6. Map of the central coast of Texas showing the extensive fluvial plain deposited during a Pleistocene low stand of sea level. The heavy lines mark former river channels. Modified from McGowen, J. H. et al., 1976, Environmental Geologic Atlas of the Texas Coastal Zone: Bay City—Freeport Area, Austin, University of Texas Bureau of Economic Geology, p. 12.

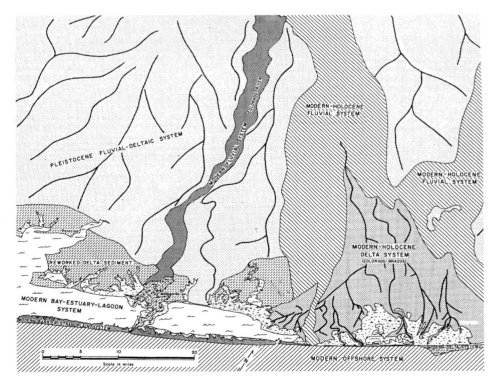

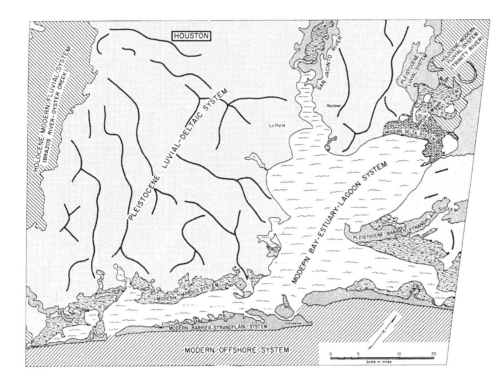

Figure 3-7. Map of the Texas coast between Houston and Galveston showing an extensive Pleistocene delta system with numerous distributary channels (dark lines). Modified from Fisher, W. L. et al., 1972, Environmental Geologic Atlas of the Texas Coastal Zone: Galveston—Houston Area. *Austin, University of Texas, Bureau of Economic Geology, p. 12.*

during the Pleistocene produced extensive fluvial plain deposition with a combination of sand and mud. Most of the sand accumulated in the form of channel deposits with adjacent muddy flood plain deposits (Fig. 3-6).

Many of the rivers carried so much sediment and had such a large discharge of both water and sediment that they produced deltas. These river deltas contained sand in their distributary channels but were dominated by mud (Fig. 3-7). Many of the rivers emptied into large, shallow estuaries similar to those present along the Gulf Coast today. These deltas are called *bayhead deltas* due to their development in the estuaries, a type of coastal bay. They grew gulfward and in some cases completely filled the estuary. Some of these deltas were deposited by the ancestral relatives of the present rivers such as the Colorado, Brazos, and Trinity of Texas, the Mobile of Alabama, the Appalachicola in the Florida Panhandle, and others. The present Guadalupe River delta in San Antonio Bay on the Texas coast is an example of what these bayhead deltas look like (Fig. 3-8).

Because of the size of its drainage basin and the volume of discharge,

*Figure 3-8a. Schematic
diagrams showing
the development of an
estuarine/barrier system
over the past 10,000
years. The final example
includes a bayhead delta
at the mouth of the river
that empties into the
estuary.*

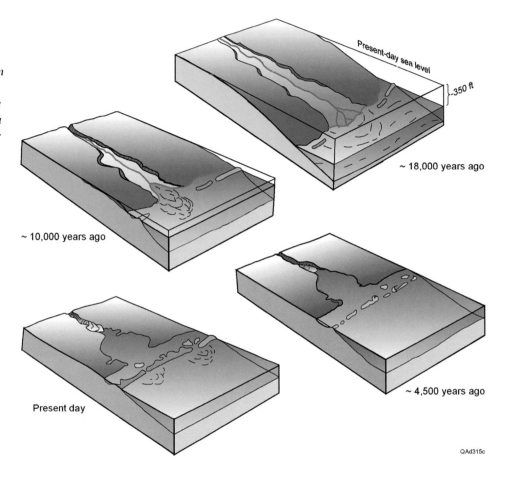

the ancestral Mississippi River was in a class by itself. Its drainage basin
included most of the area covered by the extensive ice sheets and the meltwater carried huge amounts of *outwash* sediment that accumulated as the
ice sheets melted in the upper Midwest. The ancestral Mississippi changed
from a meandering stream to a braided stream as excessive sediment
choked its channels. Remnants of these braided deposits are preserved in
northern areas of Louisiana and western Mississippi.

These extensive Pleistocene coastal deposits partially covered the previ-
ously deposited Pliocene deposits such as the Goliad Formation in Texas, the
Citronelle Formation east of the Mississippi Delta, and the carbonate Tamiami
Formation in south Florida (Table 3-2). The Goliad and Citronelle formations

2 km

Figure 3-8b. Aerial photograph showing the Guadalupe River delta as it develops in San Antonio Bay, Texas, a good example of a bayhead delta. Courtesy U.S. Department of Agriculture.

were deposited primarily by braided streams whereas the Tamiami Formation has a significant marine carbonate character. On the Texas Coast, the Lissie Formation is older Pleistocene fluvial deposits and the adjacent Beaumont Formation is more clayey due to its fluvial-deltaic character (Fig. 3-9).

The coasts of the Yucatan Peninsula and Florida were quite different from the rest of the Gulf because they are carbonate platforms upon which river systems do not develop well. The Yucatan is formed entirely

TABLE 3-2. Quaternary stratigraphic units on the Texas and Louisiana Gulf coast.

Series	Glaciation and Interglaciation	Texas	Louisiana
Holocene		unnamed	unnamed
Pleistocene	Wisconsinan	Beaumont Fm*	Prairie Fm
	Illinoisan/Sangamon	Lissie Fm	Montgomery Fm
	Kansan/Yarmouth	Lissie Fm	Bentley Fm
	Nebraskan/Aftonian	Willis Fm	
			Williana Fm
Pliocene		Goliad Fm	Goliad Fm

*Fm = formation

Figure 3-9. Map of a portion of the south Texas coast near Raymondville showing the presence of Pliocene deposits landward of the Pleistocene deposits. From Brown Jr., L. F. et al, 1980, Environmental Geologic Atlas of the Texas Coastal Zone: Brownsville—Harlingen Area. Austin, University of Texas, Bureau of Economic Geology, p. 49.

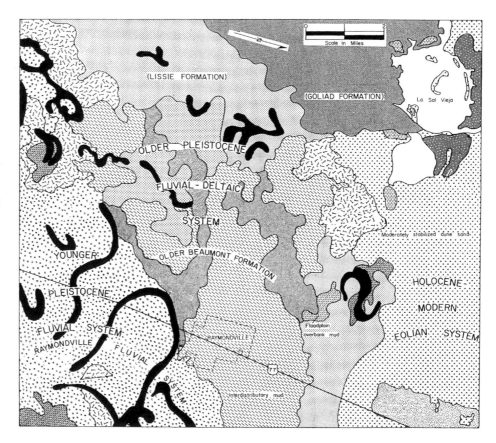

of carbonates because, during its history, there has not been a source or a mechanism for sand and mud to be transported to it. Consequently, the coast is characterized by carbonate material only, much like the Bahamas.

During the Miocene Epoch, sediments from the southern Appalachians filled the wide straits between what is now Georgia and Florida and permitted eroded sand and mud to be transported over the Florida Platform (Fig. 3-10). These sediments cover the peninsula with a veneer of siliciclastic sand. Large river systems were not present here, partly because of the limited land mass and partly because carbonate platforms tend to develop karst surfaces with sinkholes, caves, and disappearing or underground streams. Consequently there was little sediment carried by rivers to the coast. The result is an absence of fluvial plains and deltas along the Florida Gulf Coast. Pleistocene deposits like the Tamiami Formation are dominated by carbonate shell debris.

The Gulf Coast of Mexico exhibits a Plio-Pleistocene complex that is

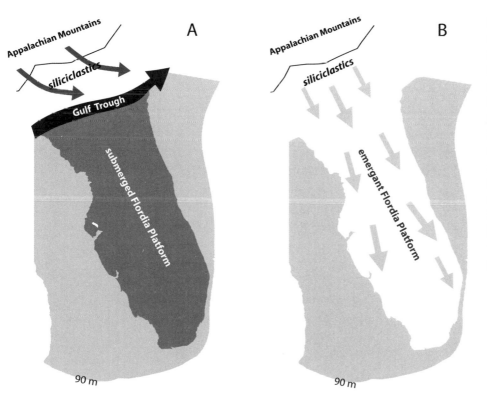

Figure 3-10. General maps of the Florida peninsula a) before it was attached to the continental United States and b) after the filling of the Straits of Florida which permitted sediment eroded from the Appalachian Mountains to be transported to form a veneer over the Florida Platform. Modified from Florida Geological Survey, www.dep.state.fl.us/geology.

Figure 3-11. Map of the southern Gulf Coast of Mexico west of the Yucatan Peninsula showing the terraces of the Mesozoic and Cenozoic ages and the modern barrier system. From West, R. C. et al., 1969, The Tobasco Lowlands of Southeastern Mexico, Technical Report 70, Baton Rouge, La:, Coastal Studies Inst., Louisiana State University, p. 34.

similar to other parts of the Gulf but has received much less attention from researchers. A coastal reach of about 250 km to the west of the Yucatan Peninsula was studied by researchers from Louisiana State University in the early 1960s. Their work shows late Tertiary highlands that are fronted by Pleistocene terraces. These terraces extend to within about 25 km of the present shoreline. Between the Pleistocene terraces and the modern coastal barrier system is a plain of fluvial-deltaic deposits supplied by three rivers that emanated from the highlands (Fig. 3-11).

During sea-level rise when ice sheets were melting, a variety of depositional environments developed. Space for sediment accumulation increased but the amount of sediment being carried to the coast decreased because base level, the position below which erosion does not take place, fell. Pleistocene sediments that had accumulated on the extended coastal plain during low sea level were drowned as sea level rose. As the shoreline

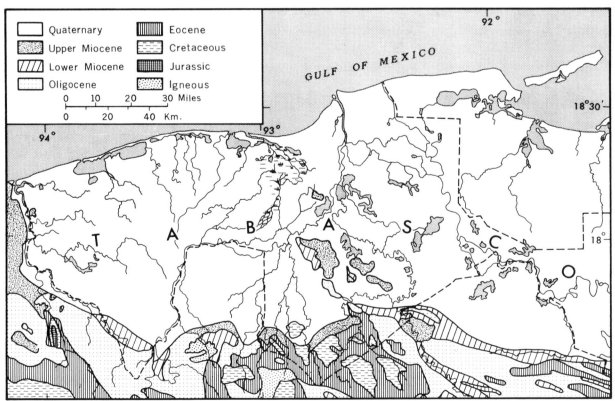

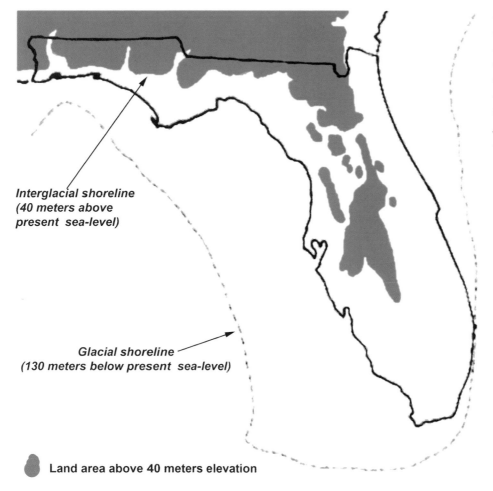

**Interglacial shoreline
(40 meters above
present sea-level)**

**Glacial shoreline
(130 meters below present sea-level)**

Land area above 40 meters elevation

Figure 3-12. Map of the Florida peninsula showing the position of the shoreline at the lowest and highest position of sea level during the Quaternary Period. Modified from Florida Geological Survey, www.dep.state.fl .us/geology.

advanced, waves reworked the Pleistocene sediment and it became enriched in sand. The mud was temporarily suspended and was transported seaward by regional currents and wave-generated currents.

The sea-level highstands that took place during isotopic stages 5–11 (see Fig. 3-3b) produced depositional units that generally correspond to those high-stand positions. In Florida, for example, the highest shoreline trend parallels the length of the peninsula where upland deposits are dominated by coastal sand dunes. By contrast, the lowest position of sea level during the Pleistocene exposes an area nearly twice the size of the present Florida Peninsula (Fig. 3-12).

Along the Florida Panhandle and adjacent Alabama and Mississippi, there are paleo-shorelines shown by subtle topographic ridges (Fig. 3-13). The Mississippi River carried significant sediment to the coast throughout the Pleistocene. The rate of deposition varied depending on whether the ice sheets were melting, when the rate was very high, and when the ice sheets were advancing, as the rate was reduced.

There are wave-dominated sediment bodies that denote old shorelines that developed when sea level was relatively high and was stable for a significant period. These are composed primarily of sand and are more or less parallel to the present coastline. These sediment bodies are treated as geologic formations with names that vary with sediment composition and geographic extent. In general there are four separate regions along the Gulf: the Texas coast, Louisiana, Mississippi/Alabama/Florida Panhandle, and the Florida Peninsula (Fig. 3-13). The basic situation along most of the northern Gulf Coast is that there are late Pleistocene fluvial deltaic deposits underlying or up-dip (landward) of the sandy barrier-strand plain deposits. These sandy deposits tend to present themselves as terraces.

Along the Texas coast the Ingleside Formation overlays the Beaumont Formation, an older muddy unit that represents fluvio-deltaic environments. This formation extends along the entire Texas outer coastal plain. The Ingleside terrace is a Stage 5e accumulation that is about 125,000 years old (Fig. 3-14). It accumulated at a time at which sea level was 3–5 m above present. The name comes from the town of Ingleside on the north shore of Corpus Christi Bay. It is common for these barrier-strand plain deposits to be capped by dunes. As a result, even though sea level was up to 5 m above present, the surface of the terrace is commonly at a higher elevation.

In Louisiana, geomorphic terraces similar to the Ingleside Terrace are represented by the Willis Formation overlain by the Prairie Formation of Pleistocene age. The northeastern Gulf Coast has the Citronelle Formation, a similar unit that forms the Pamlico Terrace. On the Florida Peninsula, the term Pamlico is also used to refer to the position of sea level and the shoreline but the deposit is the Tamiami Formation, a largely carbonate unit with considerable amounts of shell hash.

Some similar, but typically smaller, shore-parallel sand bodies are underwater. Once there was a restarting of significant melting of the ice sheets, sea level rises again, sometimes rather rapidly. If the rate of rise is

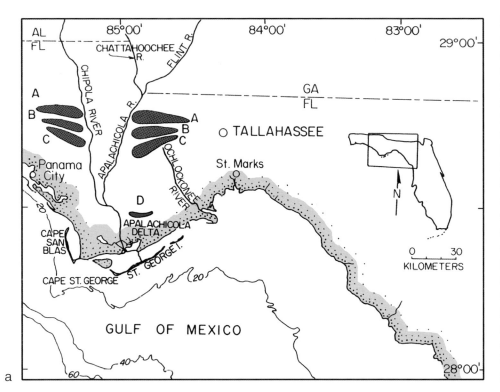

Figure 3-13. (a) Sand ridges representing old shorelines along the northeastern coast of the Gulf of Mexico (from Donoghue, J. F. and W. F. Tanner, 1992, Quaternary terraces and shorelines of the panhandle Florida region. SEPM Spec. Publ. 42, p. 234) and (b) cross section across northern Gulf showing terraces and old coastal sand ridges. From Otvos, E. G., 2005, Numerical chronology of Pleistocene coastal plain development: extensive aggradation during glacial low sea level. Quaternary International, 135:95–113.

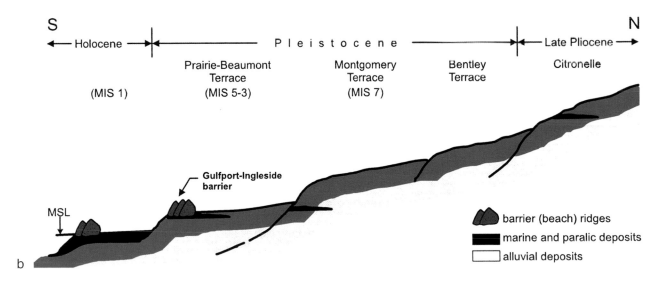

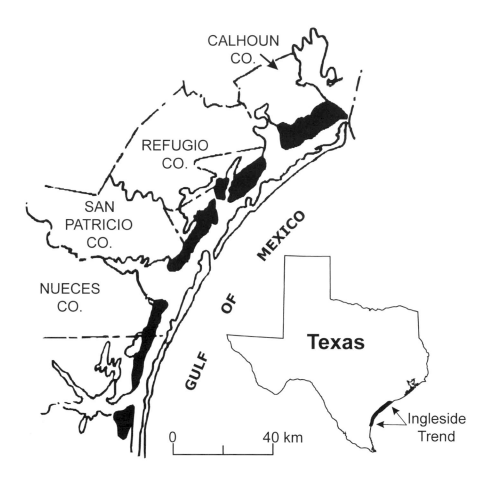

Figure 3-14. Map of the Texas coast showing the extent of the Ingleside terrace (dark band) that was deposited during isotope Stage 5e. From Otvos, E. G. and M. J. Giardino, 2004, Interlinked barrier chain and delta lobe; northern Gulf of Mexico. Sedimentary Geology, *169:47–73.*

about one centimeter or more per year, these shore-parallel sand bodies will be passed over by the rising water and not destroyed because there will not be enough time for the coastal waves to redistribute the sand. There are locations on the present continental shelf of the Gulf where relict sand bodies like these have been preserved. These channel accumulations are identified by both detailed bathymetric surveys and high-resolution seismic surveys (Fig. 3-15). Some are at the surface of the present continental shelf and others are buried or partially buried by more recent sediments.

There is also other evidence of stable sea-level conditions. Hermatypic corals are restricted to shallow water and develop reefs like the modern reefs in the Florida Keys and along the Yucatan Peninsula. Remnants of

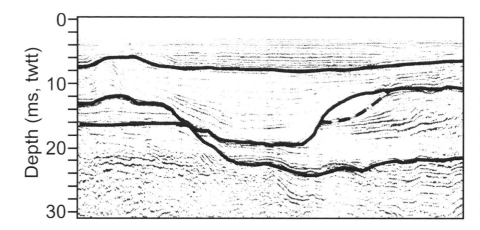

Figure 3-15. High-resolution seismic cross section from the continental shelf near Galveston showing an old river channel. From Simms, A. R. et al, 2008, Mechanisms controlling environmental change within an estuary, Corpus Christi Bay, Texas USA. Geological Society of America, Special Paper 443:121–46.

former reef environments are present near the 100 m contour on the Texas (Fig. 3-16) and Florida shelves. These relict reefs record the shallow, nearshore environments of lower stands of sea level.

The chronology of the advances and retreats of sea level has been recorded in the stratigraphy and geomorphology of the outer coastal plain. The older Pleistocene deposits accumulated on what is now the outer coastal plain. Along most of the Gulf Coast these Pleistocene accumulations are now cultivated as cotton and sorghum fields, rice paddies, and pastures for grazing cattle. In some areas, such as the northeastern Gulf, there is considerable timberland. In Florida there is a combination of pastures and piney woods.

Going gulfward, below present sea level, are some of the same sediment patterns and depositional environments. When sea level was low the same depositional environments that are now present on the outer coastal plain were accumulating on the exposed continental shelf that was then the outer coastal plain. Professor John Anderson and his students at Rice University have been studying these Pleistocene depositional environments on the continental shelf using high resolution seismic surveys. Such investigations provide cross-sections that show channels, layers of sediment, and stratigraphic discontinuities. By combining many seismic lines, these surveys permit reconstruction of maps showing the distribution of these older depositional environments (Fig. 3-17).

Glaciation caused changes in climate along the Gulf of Mexico even

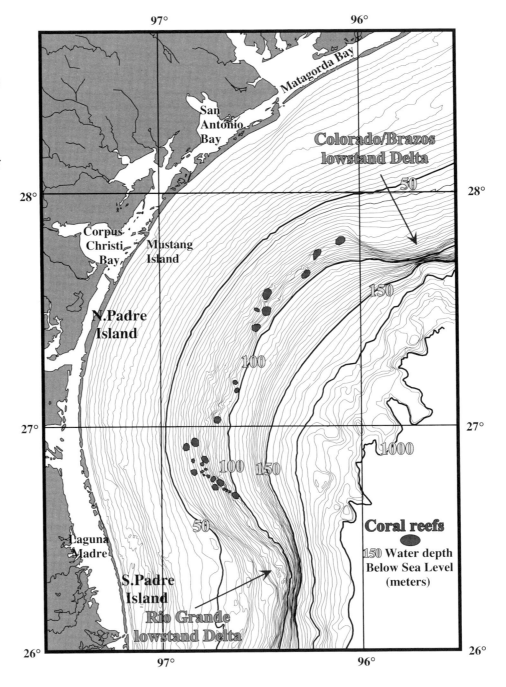

Figure 3-16. Map of the Texas continental shelf showing the position of old coral reefs, an indicator of old sea level and shoreline location. Modified from Berryhill, H. L. 1978, Late Quaternary Facies and Structure, Northern Gulf of Mexico, *Studies in Geology No. 23, Tulsa, Oklahoma: American Association Petroleum Geologists, p. 20.*

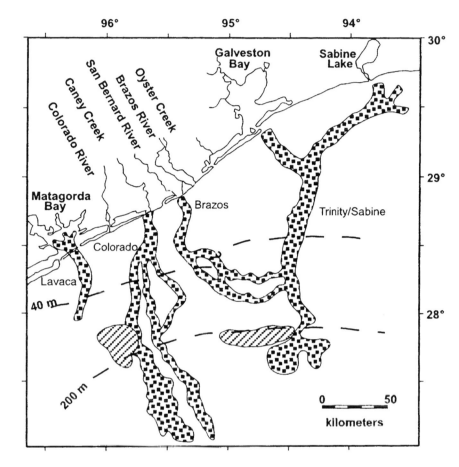

Figure 3-17. Map showing a reconstruction of Pleistocene depositional environments in the area of the continental shelf that was exposed between 80,000 and 22,000 years before present when sea level was much lower than the present time. Modified from Abdullah, K. C., J. B. Anderson, J. N. Snow, and L. Holdford-Jack, 2004. The Late Quaternary Brazos and Colorado Deltas, offshore Texas, U.S.A.—their evolution and the factors that controlled their deposition. In, Late Quaternary Stratigraphic Evolution of the Northern Gulf of Mexico Margin, ed. Anderson, J. B. and R. H. Fillon, 237–69. SEPM Special Publication 79, Tulsa, Oklahoma: SEPM, The Society for Sedimentary Geology.

though the southernmost edge of the ice sheet was more that 1000 km to the north. Rainfall was significantly less than at present. For example, in Louisiana annual rainfall was about 100 cm, which is less than at present. The Florida Peninsula and South Texas were essentially deserts with scrub vegetation and dunes (Fig. 3-18). The Gulf coastal plain contained forests similar to what are now found in the upper Midwest.

Summary

The Quaternary Period was the time in earth history when there were both rapid and major changes in sea level due primarily to the growth

Figure 3-18. General map of North America showing ice sheets and the locations of deserts during sea level lowstands in the late Pleistocene. Modified after Goudie, A., 1983, Dust storms in space and time. Physical Geography, 7:502–30.

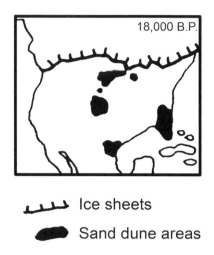

⌐⌐⌐⌐ Ice sheets

◼ Sand dune areas

and melting of extensive ice sheets. Falling sea level exposed essentially all the present continental shelf to the atmosphere and typical surface processes. River systems extended across the shelf and typically dumped their sediment lobe down the continental slope. The only deltas on the shelf were relatively small and formed during stillstands or times of very slowly rising sea level. As sea level rose over the present shelf it also reworked previously deposited fluvial sediments. Stillstands or slow rises in sea level permitted barrier islands to form, some of which still exist as relict sand bodies. Only the most recent fall and rise of sea level can be well interpreted because it is the most accessible to investigation.

Important Readings

Anderson, J. B. and R. Fillon, (eds.). 2008. *Response of Upper Gulf Coast Estuaries to Holocene Change and Sea-Level Rise.* Special Paper 443. Boulder, Colo.: Geological Society of America. A compendium of papers, mostly by Anderson and his students at Rice University, on the stratigraphy of the shallow shelf sequences.

University of Texas, Bureau of Economic Geology, various authors, 1973–80, *Texas Coastal Zone Atlas.* This is the landmark 7-volume work on the depositional sedimentary environments of the Texas Gulf Coast is organized into regional volumes from the northeastern coast (Beaumont-Port Arthur area, 1973) to the south (Harlingen-Brownsville area, 1980). Other states followed this lead.

4

Melting Ice Sheets and Sea-Level Rise

The maximum glaciation of the late Wisconsinan stage of the Pleistocene Epoch occurred about 20,000–22,000 years ago. As these ice sheets melted, sea level in the Gulf of Mexico and throughout the world began to rise from a low of about -130 m relative to its present position (Fig. 4-1). The rate and direction of sea-level change from initiation of the rise 18,000 years ago until the present was quite complicated. In fact, the position of the maximum lowstand of sea level was different in various parts of the world. Some data show that in the Gulf of Mexico sea level was not as low at this time as it was in much of the rest of the world. This is likely due to the combination of subsidence caused by the mass of the ice on North America and the compaction of the sediments associated with the Mississippi Embayment.

It is appropriate at this point in the discussion to consider the time scales. The time over which sea level has risen in the Gulf of Mexico from this all-time low to its present level began before people migrated into North America. In fact, this time of low sea level provided the conditions under which it was possible for people to cross from Asia to Alaska about 15,000 years ago across the Beringia land bridge. They eventually migrated down across the North American continent and probably first reached the Gulf of Mexico about 12,000 years ago.

The rapid sea-level rise that took place at the end of the last glaciation caused the shoreline to move rather quickly across the continental shelf. The early portion of this period of sea-level rise was the latest Pleistocene and the rest occurred in the Holocene Epoch. The Holocene is a time period named by paleontologists, scientists who study fossils. They have designated the period from 10,000 years to the present as Holocene because all remains of organisms older than that are considered fossils. This is an entirely artificial designation and has no significance in terms of earth processes or the genesis of geologic features. This explains why this trend in sea-level rise is continuous without regard for the Pleistocene-Holocene boundary.

Figure 4-1. Diagrammatic map of the reconstruction of the surface of the continental shelf at the beginning of the last period of sea level rise about 18,000 years ago. From U.S. Environmental Protection Agency, www.epa.gov/.

There are thousands of radiometric dates, primarily carbon-14 (C^{14}), from a wide variety of samples and locations across the continental shelf of the Gulf of Mexico. Typically these dates are from salt marsh and mangrove peats, oysters, the coquina clam (*Donax*), and beachrock. All were originally at or very near sea level when they were alive. These data provide information that can be plotted as time and depth parameters to construct a sea-level curve. Obviously each curve, regardless of where it is applied, has a scatter of data points. This scatter is partly due to location during the time the biogenic material was alive, the fact that it might have been moved a bit, the date determined has some error, and there might have been some contamination of the material. Nevertheless, it is possible to construct a curve that accurately shows the trend in sea-level rise as the ice sheets melted.

We can make a general summary. There was an overall rapid rise of nearly a centimeter per year from 18,000 years ago until about 6000–7000 years ago. There were numerous perturbations during this time including short periods of sea-level fall. The overall average rate of sea-level rise then slowed to about 2 mm per year until about 3500 years ago. From that time until the present there are multiple possibilities and they may change from one coastal region to another: 1) sea level may have reached its present position by ~3500 years ago and remained essentially stable;

2) sea level may have been rising very slowly from 3500 years ago to the present with few significant changes in rate or direction; or 3) sea level may have risen and fallen a meter or so above and below the present position. The last interpretation has been made by researchers who study and date beach ridges along the present coast.

Around the world there are many versions of the sea-level curve that show the rise as Pleistocene glaciers melted. Even in the Gulf of Mexico area there are differences. There are, however, similarities in that a three-part curve with different and decreasing average rates of rise tends to be common. A curve recently produced by M. Toscano and I. MacIntyre of the Smithsonian Institution shows this pattern well. The initial and steep section of sea-level rise averages more than 5 mm per year up to about 7000 years ago. A slowing to about 1.5 mm/yr took place from then until about 3000 years ago when the rate slowed again (Fig. 4-2).

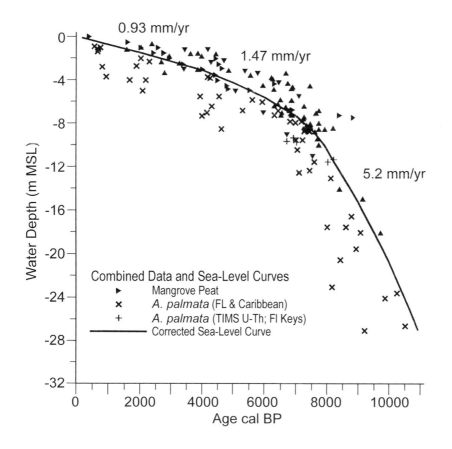

Figure 4-2. Sea level curve that shows the three distinct periods and rates of rise since the time that extensive glacial melting began. From Toscano, M. A. and I. G. MacIntyre, 2003. Corrected western Atlantic sea-level curve for the last 11,000 years based on calibrated dates ^{14}C dates from Acropora palmata framework and intertidal mangrove peat. Coral Reefs, 22:257–70.

The Florida Peninsula is the best location to measure sea-level position because it is tectonically stable and there is no subsidence, a characteristic of carbonate platforms. The same can be said for the Yucatan Peninsula but there has not been nearly as much research on this relatively remote region. Recent work at the Florida Geological Survey has produced a sea-level curve based on thousands of dates. This curve shows the typical steep initial slope that slowed gradually at about 7000–6000 years ago (Fig. 4-3). The last few thousand years have been represented in various ways on the Florida coast. The most recent plot for this period shows that sea level changed from about -1 m to +1 m multiple times (Fig. 4-4). Other plots for this same period show a gradual rise in sea level using data collected from the southwest Florida coast.

The sea-level curve for the same time frame for the Texas coast shows a similar but slightly different pattern. Because of thick sand and mud sequences on the continental shelf and adjacent coast, compaction and subsidence play a role in this curve. A recent version from the group at Rice University shows considerable detail. This curve has been constructed using peat and articulated samples of the coquina clam (*Donax*), both excellent indicators of sea level. These investigators have been able

Figure 4-3. Sea level curve developed from data taken on the Florida Platform which is the most stable part of the U.S. Gulf of Mexico Coast. From Balsillie, J. H. and J. F. Donoghue, 2004, High resolution sea-level history for the Gulf of Mexico since the last glacial maximum, Report of Investigations No. 103, Tallahassee, Florida Geological Survey.

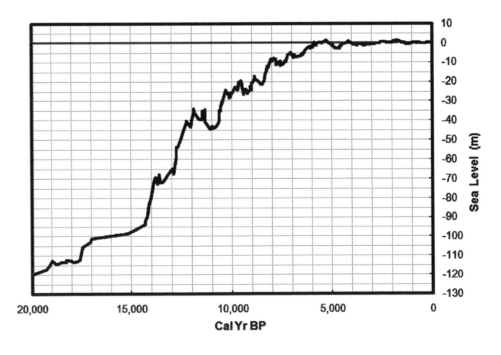

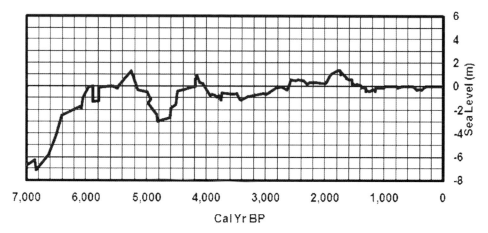

Figure 4-4. Sea level curve for the late Holocene on the Florida coast showing short periods of sea level being higher and lower than the present position. These periods of high sea level are not observed throughout the Gulf Coast. From Balsillie, J. H. and J. F. Donoghue, 2004, High resolution sea-level history for the Gulf of Mexico since the last glacial maximum, Report of Investigations No. 103, Tallahassee, Florida Geological Survey.

to show periods of slow and rapid sea-level rise within the general trend previously described for the last 10,000 years. These are at 8700, 8200, and 7600 years before present (Fig. 4-5).

There have been recent reports by Dr. Michael Blum and colleagues of sea-level highstands along the Texas coast during the Holocene. From multiple sites he interprets sea level to have been 1–2 m above its present position about 6000 years ago. Sandy and shelly deposits that are presently higher than sea level are considered to be old beach deposits where the uprush and backwash from waves took place. Many researchers are not convinced that these sediments and strata interpreted to be indicators of sea level are, in fact, properly interpreted. The alternative interpretation of these deposits is that they are the result of storms. More work is needed on possible Holocene highstand positions of sea level in order to resolve this conflict of ideas.

Continental Shelf

Consider that the average rate of sea-level rise was about 1 cm per year during the first two-thirds of the most recent 18,000 year period of sea-level rise. That would be 1 m per century or about 2 ft in a typical lifetime. This rate of sea-level rise is about the maximum during the post-glacial period. The present rate of sea-level rise is about 25% of that maximum rate or less than 30 cm (1 ft) per century. These rates must be kept in mind as sea-level rise during historical times is discussed.

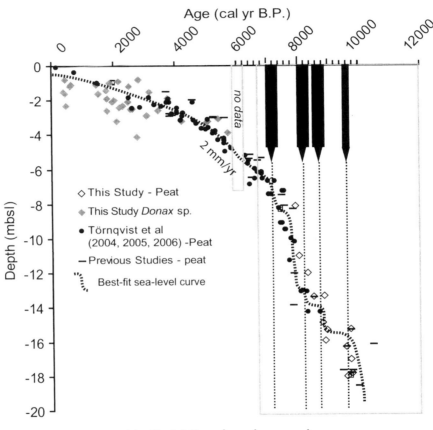

Figure 4-5. Sea level curve developed from data taken from the northern Gulf Coast. Note the short periods of sea level stability followed by rapid rises. From Milliken, K. T. et al., 2008, A new composite Holocene sea-level curve for the northern Gulf of Mexico. Geological Society of America. Special Paper 443:1–12.

NoGoM Sea-level record

The surface over which the sea transgressed was the previous outer coastal plain that developed over about 75,000 years during which the Wisconsinan-aged glaciers developed. During this lowstand, abundant sediment was transported to the coast. This Wisconsinan coastal plain was similar to the present outer Gulf coastal plain which is characterized by a combination of deltaic and fluvial depositional systems with barrier strand plains. Relief was low with fluvial channels and flood plains, and delta plains containing marshes, interdistributary plains, and channels (Fig. 4-6).

Because sea level was rising rapidly and the shoreline was moving over the outer portion of what is now the continental shelf, there is little information preserved on the resulting features. The drowning of that coast took place with nominal reworking of previous older sediments. This pro-

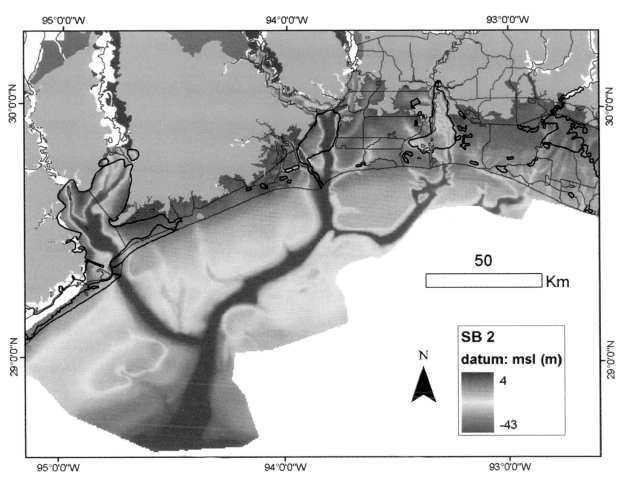

Figure 4-6. Inner shelf showing the typical fluvio-deltaic complex that formed during the Pleistocene as glaciers advanced and sea level was falling. Courtesy of K. T. Milliken.

vided a thin veneer of sandy mud overlying the older Pleistocene sediments deposited while sea level was falling during the Wisconsinan stage.

Sea level rose as river discharge increased from the melting ice sheets. This rise in sea level caused waves and tides, along with the currents that they generate, to rework the surface of the outer coastal plain. This reworking of thc late Pleistocene sediments resulted in a concentration of coarser sediments, typically sand, along the moving shoreline. Fine sediments were carried further gulfward and in some cases during the early stages of the rise cycle, far enough to eventually cascade down the continental shelf. Eventually a blanket of muddy sediment was draped over much of the continental shelf.

Figure 4-7. Oblique aerial photograph of a coastal plain river showing sandy point bars in a meandering river.

While the rise in sea level took place, the rivers of the coastal plain were discharging water from the melting ice sheets at enormous rates but sediment discharge rates continually decreased. This was because as river gradients decreased, rivers become increasingly meandering. Low gradient rivers accumulate a great deal of sand on point bars at the meander loops (Fig. 4-7). Mud is deposited on adjacent flood plains during periods of high discharge, or overbank conditions (Fig. 4-8). Oxbow lakes were widespread as meander loops were cut off (Fig. 4-9).

The changes accompanying sea-level rise during the last 18,000 years were not different than those that had taken place during previous episodes of Pleistocene glacial advance and retreat. The controlling factors in the specifics of these changes were in the rates and directions of sea-level change. The overall change was an increase in sea level of about 130 m but there were many perturbations within that 18,000 year period.

We do not know a lot about the Pleistocene sediment bodies and geomorphology beneath the modern muddy sediment layer. Detailed, high resolution surveys and coring have been conducted by the petroleum industry on this outer part of the continental shelf but most of these data are not available for academic research. Similar but less comprehensive

Figure 4-8. Oblique aerial photograph of a flood plain along the Mississippi River during overbank conditions. It is these overbank conditions that cause mud to be distributed across these extensive, flat areas called flood plains.

Figure 4-9. Vertical aerial photograph of the Rio Grande River as it meets the Gulf showing meander cut-offs on the modern coastal plain including oxbow lakes. Courtesy U.S. Department of Agriculture.

investigations have been conducted by academics on the inner shelf of most of the Gulf of Mexico. These studies show a range of depositional systems not unlike what we see today along the present coast, with the exception of the Florida Peninsula.

Florida Gulf Peninsula

The shelf along the Florida Peninsula is essentially sediment-starved. This limestone platform has little sediment except where old shorelines were located. There is an absence of delta and fluvial plain material, which contrasts with most of the Gulf of Mexico shelf area. This is the consequence of the absence of a significant drainage system on the carbonate platform. The limestone surface has scattered shell debris and displays karstic morphology with numerous sinkholes.

Old shorelines on the inner shelf are represented by relict barrier islands that formed during pauses in, or slowing, of the sea-level rise. These features are well-documented from the central Florida Gulf Coast by an extensive study conducted by the University of South Florida and Eckerd College faculty and students and sponsored by the U. S. Geological Survey. These old shoreline deposits are found from depths of 15–20 m and at distances of up to 20 km from the present shoreline (Fig. 4-10). Detailed surveys show that the sediment bodies that comprised these barrier islands were only about 3–4 m thick (Fig. 4-11). This is similar to some of the modern barrier islands along this coast. Radiocarbon dates from shell material within these sandy deposits show that the relict offshore barriers began to form about 7000 years ago when the rate of sea-level rise decreased. Because of the shallow depths and tropical storms and hurricanes, there has been some reworking of the linear sand bodies that made up the barriers. Even though reworking of coastal deposits has taken place during the Holocene, relict dunes are still present offshore of the modern Suwannee River delta.

Figure 4-10. Diagram from seismic profile off the coast of west-central Florida showing the presence of sand bodies interpreted to be relict barrier islands. Modified from Locker, S. D., A. C. Hine and G. R. Brooks, 2003, Regional stratigraphic framework linking continental shelf and coastal sedimentary deposits of west-central Florida. Marine Geology 200:351–78.

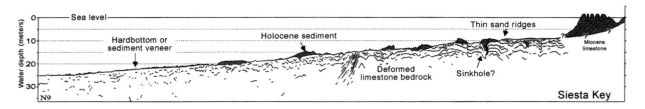

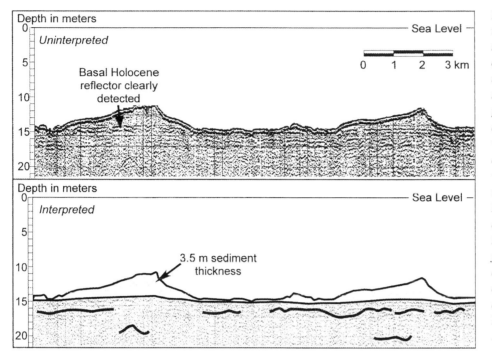

Figure 4-11. Close up of seismic profile showing sand bodies on the inner shelf off the Florida coast. These sediment bodies were probably barrier islands and were near sea level when deposited. They developed while sea level was stable or very slowly rising. Modified from Locker, S. D., A. C. Hine and G. R. Brooks, 2003, Regional stratigraphic framework linking continental shelf and coastal sedimentary deposits of west-central Florida. Marine Geology 200:351–78.

Recall that the shelf along the Florida Peninsula Gulf Coast has little sediment. There are sinkholes in the Miocene limestone that makes up most of the substrate on the continental shelf. Even though there is no significant accumulation of sediment, especially fluvial and deltaic facies, there are valleys excavated into the limestone (Fig. 4-12). These valleys are up to 40 m deep and have been filled with marine sediments of Miocene to Holocene age.

Florida Panhandle

The continental shelf adjacent to the panhandle of Florida is more like the rest of the northern Gulf of Mexico than the Florida Peninsula. This is a shelf that is underlain by terrigenous strata that derived their sediment from the southern Appalachian Mountains. We have little high-resolution seismic survey data except for an area near the present Appalachicola River delta. The group from Rice University, under the direction of John Anderson, found that a large ancestral Appalachicola delta complex occupied the Pleistocene outer coastal plain (Fig. 4-13). This huge

Figure 4-12. Diagrammatic map showing shelf valleys off the west-central Florida coast where rivers were located during low stands of sea level. Modified from Locker, S. D., A. C. Hine and G. R. Brooks, 2003, Regional stratigraphic framework linking continental shelf and coastal sedimentary deposits of west-central Florida. Marine Geology 200:351–78.

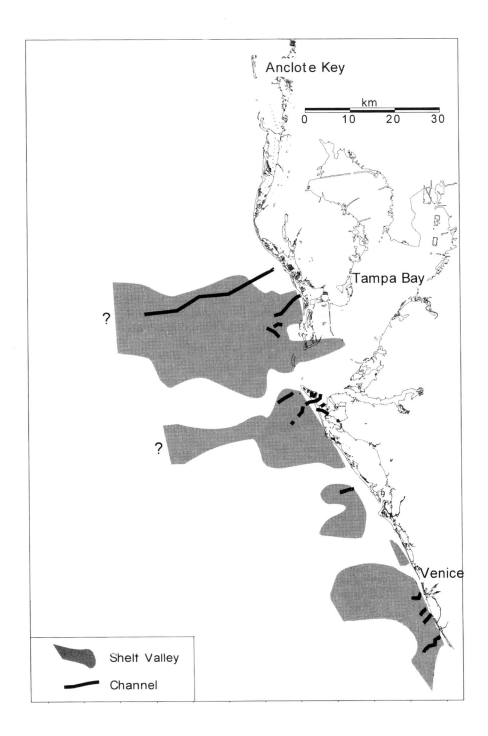

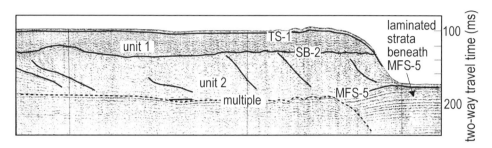

Figure 4-13. High resolution seismic section of the fluvial-deltaic complex on the continental shelf adjacent to the Appalachicola Delta on the Florida panhandle showing Gulfward progradation. This type of progradation is common in most Gulf river deltas. From McKeown, H. A., P. J. Bart, and J. B. Anderson. 2004, High-resolution stratigraphy of a sandy, ramp-type margin— Appalachicola, Florida, USA, In Late Quaternary Stratigraphic Evolution of the Northern Gulf of Mexico Margin, ed. J. B. Anderson and R. H. Fillon, 25–41, SEPM Special Publication 79, Tulsa, Oklahoma: SEPM, The Society for Sedimentary Geology.

sediment body supplied much of the sand that now forms the barrier islands along this coastal reach.

As sea level rose across this Wisconsinan-aged coastal area, fluvial-deltaic surfaces were reworked and covered mostly with mud but with sand along the shoreline. Barrier islands were not formed here until late in the sea level transgression because this shelf is much steeper than that on the carbonate platform of the Florida Peninsula.

Alabama-Mississippi

The continental shelf of Alabama and Mississippi can be considered together. There are two important areas to consider: 1) the Mobile Bay area; and 2) the Mississippi Sound area. Considerable work on this shelf has been conducted by personnel of the Alabama Geological Survey in cooperation with the U.S. Geological Survey. These studies have been undertaken to understand the Pleistocene history of this region and also to search for potential sand resources for beach nourishment projects. Location of large quantities of beach-quality sand is necessary to restore beaches eroded by tropical storms and hurricanes.

Both of these areas were covered by salt marsh and wooded communities as sea level rose and the shoreline moved across this area about 7000 years ago. Data show that the shoreline was not straight (Fig. 4-14) reflecting the broad, shallow drainage systems of the Pleistocene outer coastal plain. Data from cores taken in this area show the nature of the transgressive Holocene deposits with abundant marine shell debris overlying plant material and mud.

The geomorphology of the Mobile Bay estuary and the adjacent area offshore shows the presence of a major drainage system during the late

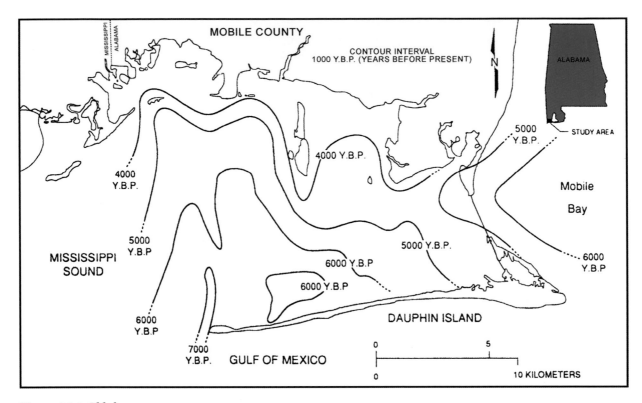

Figure 4-14. Old shorelines beneath the inner shelf of the Mississippi Sound showing numerous embayments. From Otvos, E. G., 2005, Numerical chronology of Pleistocene coastal plain and valley development; extensive aggradation during glacial low sealevels. Quaternary International, 135:91–113.

Pleistocene and the Holocene (Fig. 4-15). This drainage system was characterized by a complex system of channels that extended well out onto the shelf beyond the location of the present barrier island by the time sea-level rise slowed about 7000 years ago. Shortly thereafter, the ancestral Mobile Bay estuary was formed in the older broad valley. Sedimentation rate across the ancestral Mobile Bay was high at 9.5 cm per century. By the time that sea level reached near its present position (~4000 years ago) the extent and shape of Mobile Bay was similar to the present (Fig. 4-15). At that time, barrier islands began to form in the Mississippi Sound area (Fig. 4-16).

Louisiana Coast

The appearance of the shelf area serviced by the Mississippi River was not unlike the present. Due to the enormous sediment supply during the

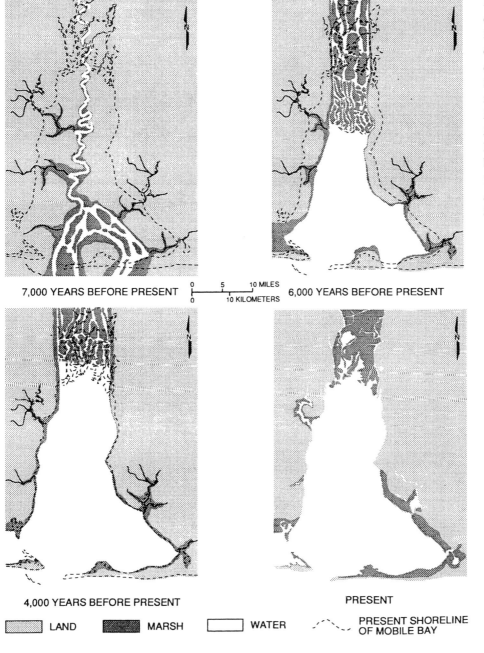

7,000 YEARS BEFORE PRESENT

6,000 YEARS BEFORE PRESENT

4,000 YEARS BEFORE PRESENT

PRESENT

0   5   10 MILES
0   10 KILOMETERS

LAND   MARSH   WATER   PRESENT SHORELINE OF MOBILE BAY

Figure 4-15. Evolution of Mobile Bay as sea level rose and flooded the existing valley eventually forming a large bayhead delta at the mouth of the Mobile River. From Hummel, R. and S. J. Parker, 1996, Holocene history of Mobile Bay, Circular C186. Tuscaloosa, Alabama: Alabama Geological Survey.

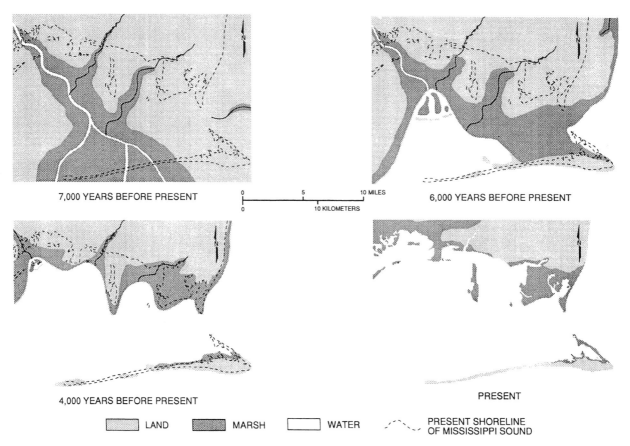

7,000 YEARS BEFORE PRESENT

0   5   10 MILES

0   10 KILOMETERS

6,000 YEARS BEFORE PRESENT

4,000 YEARS BEFORE PRESENT

PRESENT

LAND   MARSH   WATER   PRESENT SHORELINE
OF MISSISSIPPI SOUND

Figure 4-16. Evolution of Mississippi Sound as sea level rose to flood the area and begin to develop barrier islands. From Hummel, R. and S. J. Parker, 1995, Holocene history of Mississippi Sound, *Circular C185, Tuscaloosa, Alabama, Alabama Geological Survey.*

melting of ice sheets, this huge fluvial system was largely braided instead of meandering. The sediment lobe that formed the ancestral delta extended across the continental shelf. Considerable amounts of sediment were carried over the shelf edge, down the continental slope to the Mississippi Fan at depths of more than 3000 m (Fig. 4-17). This sediment accumulation is more than a kilometer thick and covers many hundreds of square kilometers. This is an enormous volume of sediment delivered by the ancestral Mississippi River.

A very thick blanket of mud covered the continental shelf adjacent to the delta. As the shoreline transgressed over this muddy surface during glacial melting, the sediment was reworked and the mud of the delta became the mud of the transgressive shelf. This scenario prevailed during the rapidly rising sea level until the slowdown that occurred about 7000 years ago. At this time, the gradient of the Mississippi River decreased and sediment supply decreased resulting in a dominantly meandering fluvial system. This system deposited its muddy discharge on the inner continental shelf in the form of lobe-shaped sediment body complexes; the Mississippi Delta. Six of these delta lobes have been identified across more than 300 km of the northern Gulf Coast (Fig. 4-18). The oldest of these began accumulating about 7500 years ago. Avulsion changed the path of the

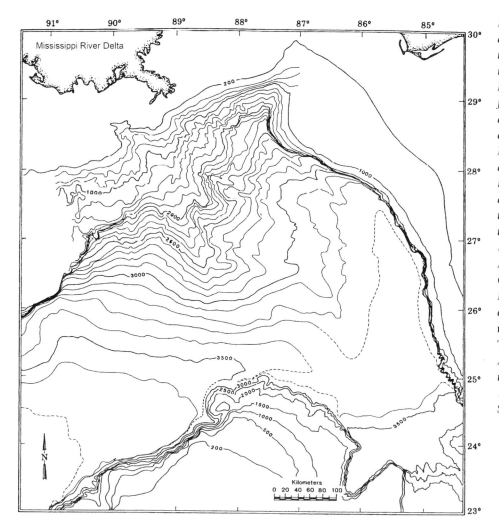

Figure 4-17. Bathymetry of outer continental margin near the modern Mississippi Delta showing the large fan-shaped sediment accumulation at the base of the continental shelf. Much of this sediment accumulated when sea level was much lower and the coast was near the edge of the continental shelf. From Bouma, A. H., J. M. Coleman, A.A. Wright-Meyer, and C. E. Stelting, 1985, Mississippi Fan, Gulf of Mexico in Submarine Fans and Related Turbidite Systems, ed. A. H. Bouma, W. R. Normark and N. E. Barnes, 1434–50, New York: Springer-Verlag.

Figure 4-18. Locations and outlines of numerous Holocene lobes of the Mississippi River delta that formed after the slowing of sea level rise about 7,000 years ago. Modified from Kolb, C. R. and J. R. van Lopik, 1966, Depositional environments of the Mississippi River deltaic plain In Deltas; Their Geologic Framework, Houston: Houston Geological Society p. 22, as it appeared in Roberts, H. H., 1997, Dynamic changes of the Holocene Mississippi River delta plain. Journal Coastal Research, 13:605–27.

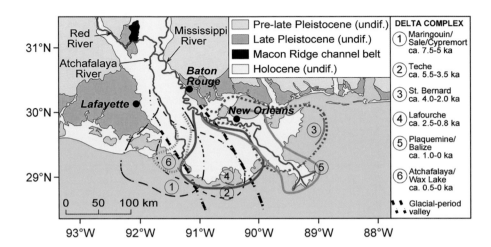

lower river and thus changed the location of the lobes that produced this huge composite delta that are all of Holocene age. Presently there are two active lobes: the older, larger from which the term "bird's foot" (Fig. 4-19) is derived, and the younger, smaller one that is actually deposited by the Atchafalaya River (Fig. 4-20), a major distributary of the Mississippi.

There are also huge, shallow sand bodies that originated as barrier is-

Figure 4-19. Satellite photo of the active portion the modern lobe of the Mississippi Delta showing the morphology that give it its "birdsfoot" designation. Courtesy USGS National Wetlands Research Center, CWP-PRA Task Force and Louisiana Department of Environmental Quality.

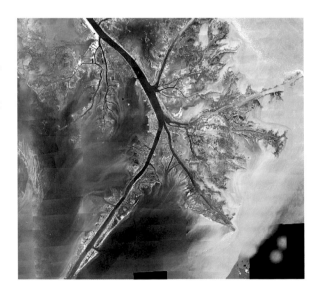

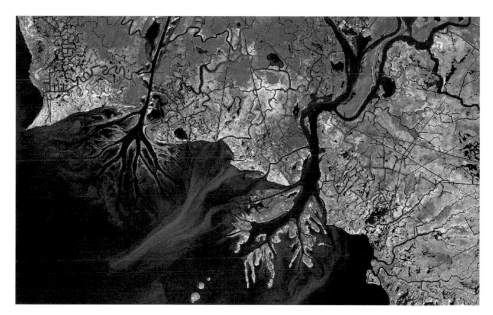

Figure 4-20. Satellite photo showing the newly formed deltas produced by the Atchafalaya River with water diverted from the Mississippi River. Courtesy USGS National Wetlands Research Center, and Louisiana Department of Environmental Quality.

lands during the slow sea-level rise of the past several thousand years. These sand bodies, such as Ship Shoal, are only a few kilometers off the present coast of western Louisiana. They are potential sources of sand for coastal protection and beach nourishment.

Texas Coast

Although there is some variation, we can discuss the Texas coast collectively. It is an extensive area that also includes westernmost Louisiana and northernmost Mexico. This 800 km long region was characterized by multiple fluvial systems that produced a Pleistocene coastal plain complex of fluvial-deltaic depositional systems. Unlike much of the Gulf continental shelf, the Texas shelf is reasonably well-known. Most of these data are the result of extensive studies by the group from Rice University. Their efforts have produced a map of this portion of the continental shelf that shows Pleistocene topography over which current topography is superimposed. The bathymetric high offshore of Freeport, Texas is interpreted as an older barrier formed during a pause in Holocene sea-level rise. During the Pleistocene there was considerable relief with several valley systems (Fig. 4-21), each of which supported a river. These rivers produced extensive and complicated fluvial-deltaic complexes, some of which extended to the edge of the shelf.

Figure 4-21. Fluvial valleys that were cut into the continental shelf off the Texas coast during periods of low sea level during the Pleistocene. From: Eckels, B. J., M. L. Fassell and J. B. Anderson, 2004, Late Quaternary evolution of the wave-storm-dominated Central Texas Shelf, In Late Quaternary Stratigraphic Evolution of the Northern Gulf of Mexico Margin, *ed. J. B. Anderson, and R. H. Fillon, 271–88. SEPM Special Publication, 79, Tulsa, Oklahoma: SEPM, The Society for Sedimentary Geology, p. 285.*

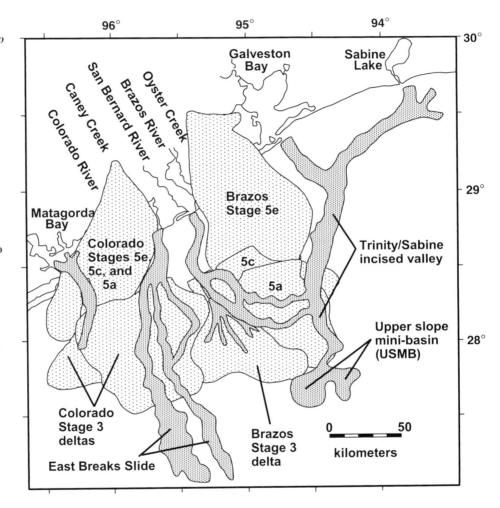

The southern portion of the Texas shelf area shows a somewhat different pattern than the northern part. The many Wisconsinan-aged fluvial depositional systems that extended beyond the location of the present coastline coalesced, flowed south, and fed the Rio Grande deltaic system (Fig. 4-22). Another aspect of this shelf was the presence of numerous coral reefs in the general location of the 100 m depth contour. These relict reefs are topographic highs and although they no longer support hermatypic corals, they are inhabited by a wide array of organisms.

The ancestral Rio Grande River system extended across the conti-

Paleo-Environment Deposition Map

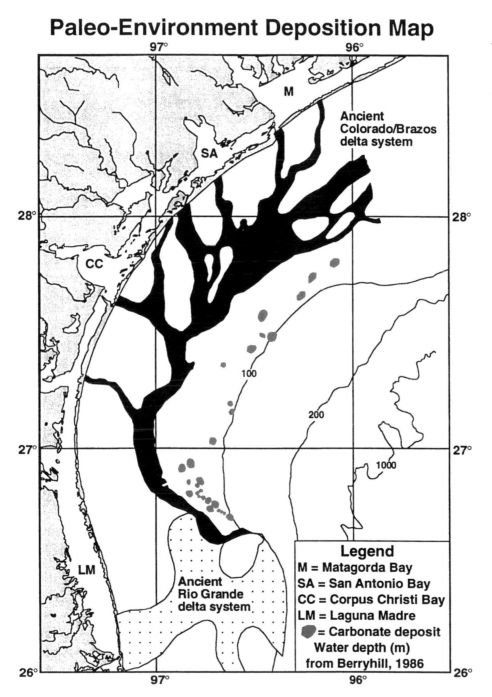

Figure 4-22. Extensive fluvial system that extends across the south Texas shelf. Note the presence of numerous coral reefs at about the middle of the shelf. From Banfield, L. A., 1998, The Later Quaternary Evolution of the Rio Grande System, Offshore South Texas. PhD. Dissertation, Houston, TX.: Rice University.

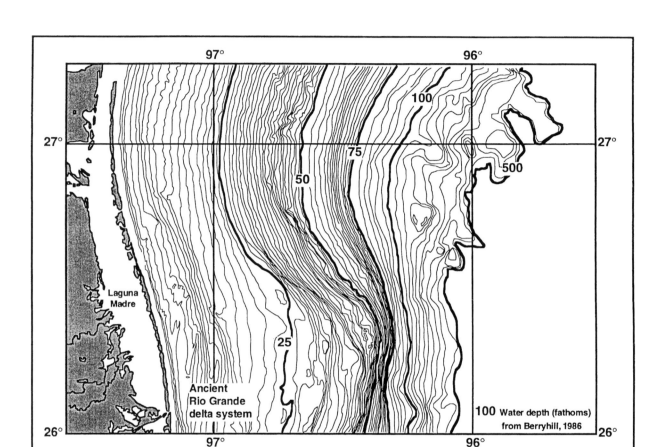

Figure 4-23. Bulge in the bathymetry showing the buried Pleistocene Rio Grande Delta system. Modified from Berryhill, H., 1978, Late Quaternary facies and structure, northern Gulf of Mexico: Interpretations from seismic data. AAPG Studies in Geology #23. Tulsa, Oklahoma: American Association of Petroleum Geologists, p. 19.

nental shelf with some sediment discharge extending to the shelf edge (Fig. 4-23). At the end of the Pleistocene the climate was more humid than it is today. There was also more rainfall, thus providing a much greater rate of sediment production to form a large deltaic complex. The advancing shoreline and adjacent surf zone reworked these deposits but left a distinct sediment bulge on the outer shelf.

Mexico Coast

We have little information about the inner shelf of Mexico during the post-glacial rise in sea level. The continental shelf between the Texas-Mexico border and the Yucatan Peninsula is the narrowest of the Gulf of Mexico and the adjacent coastal plain is similarly narrow. Well-developed drainage systems are not present so presumably the shelf does not contain extensive fluvial-deltaic deposits of Pleistocene age.

The subtidal Yucatan platform contrasts with that of the Florida platform in that it does not have a veneer of siliciclastic or terrigenous sediments. This carbonate system is limestone both above and below sea level. The late Pleistocene lowstand exposed the entire platform to karst development and there are numerous sinkholes (cenotes) across the surface. The warm, clear water conditions allowed numerous coral reefs to develop and most of these reefs are still living today.

Human Occupation

People came to the area surrounding the Gulf of Mexico several thousand years ago. When they arrived in this part of North America, the coast was many kilometers further out into the basin than at the present time. If we consider only the Holocene portion of the post-glacial sea-level rise, the shoreline was at about where the present -30 to -40 m isobath is located. There were numerous estuaries and rivers that extended across what was the outer coastal plain. These provided excellent sites for human occupation with abundant food and possibilities of easy transportation.

These sites are now under water and difficult to locate without extensive and intensive surveys. We do have, however, locations of several middens and artifacts testifying to former occupation. The middens are comprised dominantly of shells; typically oysters, other bivalves, and gastropods (snails). They are in a sense, garbage dumps of inorganic material that can last for millions of years. These middens are comprised of the shells from organisms that lived in the immediate area of occupation and that were a major portion of the protein in the diets of the early people. They may be from marine, estuarine, or freshwater animals, depending upon the location of the occupation site at the time.

Professor Nancy White of the University of South Florida has been investigating offshore occupation sites since the 1980s. She has worked with others along the U.S. Gulf Coast to summarize the known sites. Across the northern Gulf there is an identified site offshore of the present Appalachicola River delta. This site is dominated by oyster shells and has been dated at 4000 BP (years Before Present). Nearby is another site of about the same age that is composed primarily of *Rangia*, a freshwater clam. Continuing around the Gulf Coast to the Florida Peninsula there is a younger midden located near Sarasota Bay. It is 2200 years old and is composed of the shells of whelks and conchs, large gastropods.

Another extensive study on the Florida coast was conducted in the Big Bend area offshore of the present Aucilla River. The archaeology group from Florida State University headed by Michael Faught found numerous occupation sites up to almost 20 km from the present shoreline at depths of more than 4 m. They discovered about 25 sites along the paleochannel of the ancestral Aucilla River, each with various artifacts of human activity and assemblages of bones and shells. Radiometric dating of various materials in these sites indicate that the area was terrestrial until 7000 BP, coastal in 6000 BP and was inundated by 5000 BP. The oldest documented site in this area has been dated at 12,400 BP, which is probably the oldest site date from the paleo-coastal plain of the Gulf of Mexico.

There are currently investigations off the coast of Louisiana at sites that may represent human occupation. Based on accepted sea-level curves, these sites would be about 12,000 years old (Amanda Evans, personal communication). One of the oldest sites on the Texas coast has been dated at more than 8000 BP. This site was determined from cores taken as part of a Minerals Management Service (MMS) project. Human occupation along the Texas coast was widespread during the Archaic Period (7500–1500 BP). There is abundant evidence of occupation based on pottery and middens that included shells and fish bones from the beginning of the Archaic until historical time (~17th century) with gaps in the record between 6800 and 6000 years ago and from 4500 and 3000 years ago. It is possible that these were times when sea-level change was rapid and/or food was not abundant. Like the area on the eastern coast of the Gulf, sea level on the Texas coast seemed to have been stable from about 3000 years ago to the present.

Coastal Zone

For purposes of this discussion, the coastal zone includes all environments from the surf zone landward including the coastal bays. It is apparent from the overall shoreline patterns along the Gulf of Mexico that this is a drowned coast. The numerous estuaries and former estuaries that represent drowned fluvial systems provide evidence for this conclusion. There are few sites where sediment has discharged into the open Gulf at a river mouth. The most obvious is the Mississippi Delta but the Sabine, Brazos, Colorado, and Rio Grande rivers also are in that category. The majority of the open coast of the Gulf is fronted by barrier islands; most are wave-dominated but some on the Florida coast are mixed energy. Wave-dominated barriers are long and narrow with small, widely spaced tidal inlets. Mixed-energy barriers are formed by a combination of tidal currents in the large inlets and wave energy to produce these short and usually wide barriers.

The numerous rivers that feed the estuaries develop bayhead deltas, the largest of which is in Mobile Bay at the mouth of the river by the same name. There are only two areas where vegetation dominates the open coast in the entire Gulf of Mexico: the Big Bend area and the Ten-Thousand Islands area of southwest Florida. The Big Bend is an open marsh coast where several small rivers drain across to the Gulf and is essentially an open estuary. The Ten-Thousand Islands is similar in hydrography but with a mangrove-dominated coast.

The origin of the barrier islands along the Gulf Coast is definitely tied to sea level. In the 19th century there were different theories as to how barrier islands form. Two were prominent and were carried forward into the 20th century. One called upon waves to concentrate sand in the shallow water and eventually pile it up above sea level to form the beginnings of a barrier island (Fig. 4-24a). The other ascribed barriers to the drowning of beach-dune ridges as sea level rose. Coastal sediments were built into shore-parallel linear sand bodies and, when coupled with a rise in sea level, produced a barrier island with a bay on the landward side between the island and the mainland (Fig. 4-24b). In both theories, it is necessary to have a period of very slowly rising sea level or a stillstand. This gives wave action enough time to accumulate the sediment necessary to form the nucleus of a barrier island.

There now seems to be reason to choose the first theory over the second. One of the primary axioms in geology is that the "present is the key to the past." In other words, we can understand what happened in the distant past by observing what is going on at the present time. There are no known examples of situations where rapidly rising sea level drowned

Figure 4-24. Formation of a barrier island by (a) upward shoaling of sand by waves, or (b) drowning of a beach-dune ridge by rapidly rising sea level. Only the former has been observed to happen along the present Gulf Coast.

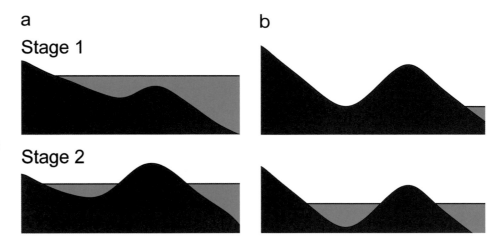

Figure 4-25. Oblique aerial photograph of a barrier island, North Bunces Key, on the Florida coast near the mouth of Tampa Bay that first became emergent in 1970. This is an example of the type of barrier island formation shown in Fig. 4-24a.

a previously existing beach-dune ridge to form a barrier island. On the other hand, there are several examples on the Gulf of Mexico coast where sand has been accumulated in shore-parallel, linear sand bodies that have developed into barrier islands (Fig. 4-25). Vegetation colonizes these sand bodies and leads to the accumulations of wind-blown sand in the form of dunes. This combination produces stable barrier islands that will persist unless a hurricane destroys them.

Florida Peninsula Coast

The most diversified area of the Gulf of Mexico coast is that of the western Florida Peninsula. As mentioned above, it has the Ten-Thousand Islands which are on the south adjacent to the Everglades, a diverse reach of barrier islands and tidal inlets, and the Big Bend, an open-coast marsh. There are numerous estuaries, both big and small, that are associated with the barrier-inlet systems.

The Ten-Thousand Islands area is a tide-dominated, mangrove coast. Studies at the University of Miami by Randall Parkinson show that the rising sea level moved over the mangrove peat until sea level reached about 2 m below the present. At that time sea-level rise slowed markedly and the coast transitioned to a prograding coast (Fig. 4-26). This was the result of

Figure 4-26. Diagrammatic profile across the Ten-Thousand Islands coast showing the transgressive and regressive phases of development that took place during the latter portion of the Holocene. Modified from Parkinson, R., 1989, Holocene evolution of the Ten-Thousand Islands area, southwest Florida. **Journal of Sedimentary Research, 59:960–72.**

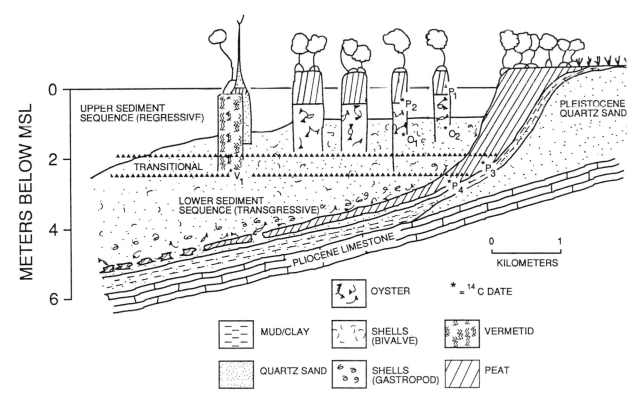

a combination of sea-level stability and a slow but steady supply of sediment from both the mainland and the Gulf.

The coastal systems of the Florida Peninsula have developed fairly recently in the period of post-glacial sea-level rise. Hundreds of radiocarbon dates have been determined from these barrier islands. The oldest barrier island, as determined by Dr. Frank Stapor and his colleagues, is Siesta Key near Sarasota which began to form only 3000 years ago (Fig. 4-27). This is a mixed-energy barrier that shows considerable progradation in the form of beach ridges. Sanibel Island, about 100 km to the south, also has a core that is 3000 years old. In addition, there are many wave-dominated barriers that are long and narrow. Anclote Key is a good example (Fig. 4-28). This island began to form only about 1500 years ago. There are also barrier islands in this system that are only decades old (Fig. 4-29).

We earlier recognized older barriers that were present on the shelf. These were small and were formed by short pauses in sea-level rise (see Figs. 4-10 and 4-11). When sea level reached near its present position about 3000–4000 years ago, there was time for the formation of substantial barrier islands. The limiting factor was the availability of sediment. This is the reason that some barriers took longer to form and that some are still in the

Figure 4-28. Aerial photograph of Anclote Key, a good example of a wave-dominated barrier island from the Florida peninsular coast.

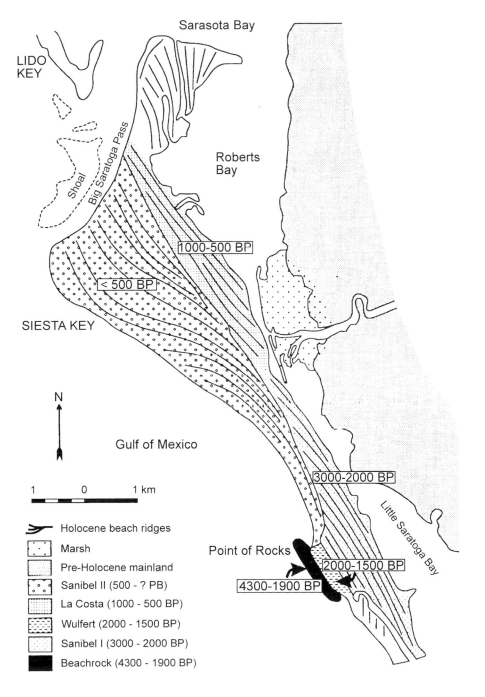

Sarasota Bay

LIDO KEY

Shoal

Big Saratoga Pass

Roberts Bay

1000-500 BP

< 500 BP

SIESTA KEY

N

Gulf of Mexico

1   0   1 km

Little Saratoga Bay

3000-2000 BP

Point of Rocks

2000-1500 BP

4300-1900 BP

Holocene beach ridges
Marsh
Pre-Holocene mainland
Sanibel II (500 - ? PB)
La Costa (1000 - 500 BP)
Wulfert (2000 - 1500 BP)
Sanibel I (3000 - 2000 BP)
Beachrock (4300 - 1900 BP)

Figure 4-27. Map of Siesta Key, one of the oldest barrier islands on the peninsular coast of Florida, showing the ages of island development. Modified from Stapor, F. W. Jr., T. D. Mathews, and J. E. Lindfors-Kearns, 1991, Barrier island progradation and Holocene sea-level history in southwest Florida. Journal of Coastal Research, 7:815–38).

Figure 4-29. Aerial photograph of Three-Rooker Bar, a young, wave-dominated barrier island adjacent to Anclote Key shown in Fig. 4-28.

process of forming. Recall that this is a sediment-starved shelf. In addition, there are no barrier islands in the Ten-Thousand Islands and Big Bend areas. The reason for this is the dearth of sediment in these areas. In both places, the inner shelf is a limestone pavement with scattered shell debris. All evidence along this coast points to a very slow rise in sea level over the past 3000 years. Because this coastal system developed on the Florida Platform which has limited drainage development, there are no significant bayhead deltas or deltas that empty onto the open shelf along this coast.

Northern Gulf Coast

The rate of sea-level rise slowed significantly about 6000–7000 years ago here and throughout the world. In fact there are several places, such as Australia, New Zealand, and southern Africa, where sea level reached its present level about that time. Worldwide, most barrier island systems are in the range of 6000–7000 years old. This is true along much of the U.S. Atlantic Coast as it is for most of the Gulf of Mexico. The younger barriers in Florida are atypical due to the general absence of sediment combined with the rapid transgression of the shoreline across this very wide and gently sloping shelf. From the Florida Panhandle west to the Mississippi Sound there are barriers that are typically 5000–7000 years old except at the Mississippi Delta. These barriers can be seen beginning to form in Figure 4-16.

Sea level in the northern Gulf Coast reached near its present position as the barriers began to form. This reach of barrier islands is complex in that some are wave-dominated (Fig. 4-30a) and others are mixed energy. They range from the long and narrow, such as Santa Rosa Island, Florida to the short and very wide, such as St. Vincent Island (Fig. 4-30b). The latter island has more than 100 beach ridges that prograde into the Gulf.

a

Figure 4-30. Vertical aerial photos showing (a) Anclote Key, a wave-dominated barrier, on the central Gulf Coast of Florida (infrared) and (b) St. Vincent Island, a mixed energy barrier along the northeastern Gulf Coast. Photograph courtesy J. F. Donoghue.

b

The differences between them are not related to any sea-level factors but to sediment supply and nearshore bathymetry.

Louisiana Coast

This area is quite different from the rest of the Gulf Coast. As sea level moved over the abandoned lobes of the ancestral delta, the lobes and the older parts of the delta were considerably modified. Here sediment had been tremendously abundant for thousands of years.

Water was very shallow and waves were small. In the western part of the delta there was reworking of sediment lobes by wave action to form what is generally called the destructive part of the delta. The result was the formation of barrier islands. These barrier islands have little elevation and this condition, coupled with compaction and subsidence, has caused sea level to overtop and destroy most of them. They also formed over a broad time span as each of the abandoned deltaic lobes was reworked. The size and location of the barriers has changed considerably during historical time (Fig. 4-31).

Figure 4-31. Barrier islands along the outer Mississippi Delta. These wave-dominated barriers were developed from the reworking by waves of the outer margin of older delta lobes. Courtesy USGS National Wetlands Research Center and Louisiana Department of Environment Quality.

Figure 4-32. Aerial photograph of Chandeleur Islands showing breach caused by Hurricane Ida in 2009. These low-lying barriers were developed from reworking of a delta lobe. Photograph courtesy Michael Miner.

As sea level moved over and flooded other abandoned lobes of the delta there was reworking and formation of other vulnerable barriers. The best example of this process is the Chandeleur Islands. These low, narrow barriers have only recently formed and are in the process of being destroyed (Fig. 4-32). These islands are overtopped and channels are cut through them by even modest storms.

Texas Coast

In this discussion the western portion of Louisiana will be included with the entire Texas Coast. The main difference between them is the result of sediment supply which is produced largely by the Mississippi River Delta complex. The Louisiana coast west of the delta and in to East Texas receives much sediment, both mud and sand. Because sea level has not moved much during the past few thousand years, there has been ample time for considerable progradation and many beach ridges have developed in western Louisiana and easternmost Texas. These are a special type of beach ridge in that they are the result of storms reworking generally muddy sediments that contain shell material. They form these special beach ridges called *cheniers* (Fig. 4-33). There was little or no change in sea level during their formation.

Moving west, the coast receives less and less sediment from the Mississippi Delta. The primary sediment supply is now from erosion of older sediments or from a few rivers that flow directly into the Gulf such as the Brazos River. There are prograding ridges in the Galveston area but these disappear further to the southwest. The remainder of the Texas coast is comprised of rather narrow wave-dominated barrier islands. These islands began to form about 5000 years ago on the central Texas coast (Fig. 4-34).

Figure 4-33. (top) Schematic map (from Byrne, J., D. O. Le Roy, and C. M., Riley, 1959, The chenier plain and its stratigraphy, southwestern Louisiana. Transactions of the Gulf Coast Association of Geological Societies, 9:237–60) (Reprinted by permission of the GCAGS), and (bottom) aerial photo of cheniers along the eastern Louisiana coast. These low shelly ridges were developed by reworking of muddy coastal sediments.

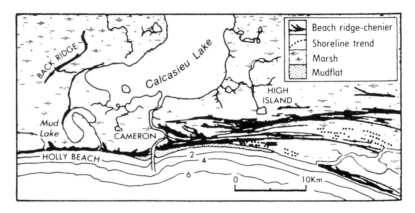

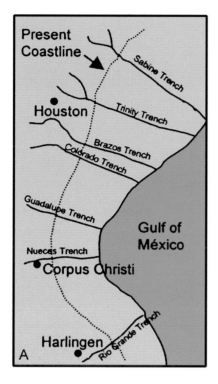
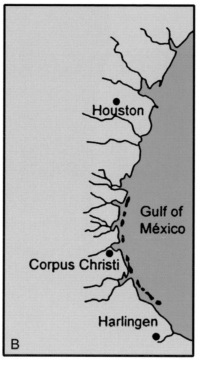
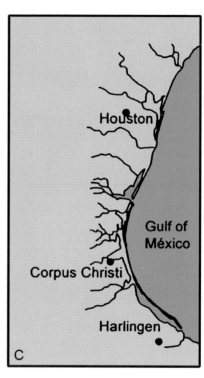

As an example of the Texas barriers we can consider Matagorda Island (Fig. 4-35). A detailed study by Bruce Wilkinson has demonstrated the development of this island and this can be translated to the other similar barriers. He found thin Pleistocene deposits preserved beneath the Holocene barriers indicating that as sea level rose during the Holocene there was reworking of the Pleistocene sediments to form the present inner shelf and barrier island complex. Other Texas barriers, particularly Galveston and Padre islands, have been investigated in detail as part of petroleum industry research.

Mexico Coast

Barrier islands continue south from Texas to the coast of Mexico. Additional barriers are present on the Tabasco coast to the western Yucatan Peninsula. Prograding beach ridges are abundant in this area (Fig. 4-36). The wide and flat coast along this part of the Gulf combined with the

Figure 4-34. The development of the Texas shoreline from the time of the glacial maximum a) through the time of the beginning of barrier island formation 6–7 thousand years ago, b) to the present shoreline, c). From LeBlanc, R. J. and Hodgson, W. D., 1959, Origin and development of the Texas shoreline. Proc. 2nd Coastal Geography Conf., Louisiana State Univ., Baton Rouge, p. 57–101. Reprinted by permission of the GCAGS.

Figure 4-35. Aerial photograph of Matagorda Island, a wave-dominated barrier on the Texas coast. Courtesy U.S. Department of Agriculture.

abundant sediment produced by the discharge of highland rivers provided the proper setting for the development of these barriers. The slowly rising sea level permitted wave action to focus at a particular shoreline location for enough time to construct the prograding beach ridges. Variation in the number and patterns of these ridges was in response to differences in sediment availability. In one area there are more than 20 beach ridges.

To the east along this coast there is no apparent change in the characteristics of the barriers as seen on the maps but their composition changes dramatically. From Isla del Carmen to the east, sediment composition is dominated by carbonate shell debris. The fluvial discharge of sediment does not extend further east to the Yucatan Peninsula. From Isla del Carmen, around the peninsula, and to the south, the coast is comprised of calcium carbonate.

Grand Cayman Island

Although not strictly on the Gulf of Mexico, the Cayman Islands, located about 200 km south of Cuba, illustrate what happened along the eastern margin of this basin. These limestone islands rest on the Caribbean crustal plate. We can learn about the late Holocene flooding of carbonate land masses from a study by Colin Woodrofe while he was at the University of Auckland in New Zealand.

Much of Grand Cayman Island is bordered by mangroves. Cores from the island and the shallow adjacent water show the presence of thick and apparently continuous mangrove peat deposits. These peat deposits are associated with lime mud (calcium carbonate) and skeletal carbonate debris from corals, mollusks, and calcareous algae. Radiocarbon dating of the upper portion of the peat layer shows that sea level has risen nearly 2 m during the past 2100 years. This rise in sea level could have come from tectonic activity, a eustatic rise in sea level, or some combination of both.

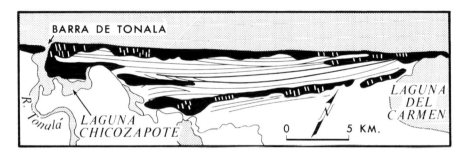

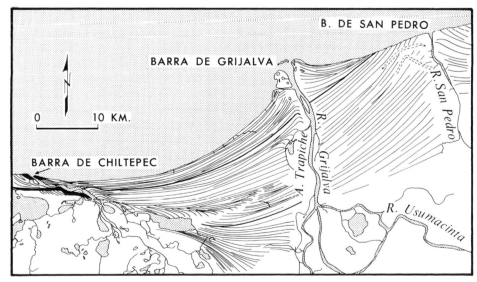

Figure 4-36. Examples of a) narrow and b) wide barriers along the Tabasco coast of Mexico. The difference is due primarily to sediment supply. From West, R.C., N. P. Psuty, and B. G. Thom. 1969, The Tabasco Lowlands of Southeastern Mexico, Technical Report. 70, Baton Rouge, Louisiana: Coastal Studies Institute, Louisiana State University.

Summary

Some generalizations about the rise in sea level over the past 18,000 years and the products that accumulated during that time period are appropriate. There were three time frames within which generalizations can be made about the rate of sea-level rise. The initial and longest period lasted until about 7000 years ago with an average rate of about 1 cm/year. There were times when sea-level rise was slower and even times when sea level was falling during that period because ice sheets grew for a few centuries. From about 7000 to 3500 years ago, the rate of sea-level rise slowed markedly in the Gulf of Mexico to about 3 mm/year. There are some researchers who even think that it rose above present sea level during this time. There are various theories about the position of sea level during the past few thousand years: 1) it was essentially stable at its present level; 2) it rose very slowly; or 3) it moved above and below its present level. The resulting products included estuarine and lagoonal deposits on the shelf along with sinuous, fluvial channel deposits. Most present barrier islands formed between 3000 and 7000 years ago.

Important Readings

Anderson, J. B. 2007. *The Formation and Future of the Upper Texas Coast.* College Station: Texas A&M Press. A beautifully illustrated treatise that includes all geologic aspects of the origin of the present situation on this part of the Texas coast.

Anderson, J. B., and R. H. Fillon, (eds.). 1996. *Late Quaternary Stratigraphic Evolution of the Northern Gulf of Mexico.* SEPM Special Publication No. 79. Tulsa, Okla: The Society for Sedimentary Geology. The development of the northern coast of the Gulf based on detailed research projects, primarily by the team from Rice University.

National Research Council (NRC). 1990. *Sea Level Change.* Geophysics Study Committee. Washington, D.C.: National Academy of Science. One of the several volumes devoted to sea level and climate problems from the NRC using noted experts from many institutions.

5

What Is Happening Now?

The previous chapter concentrated on the development of coastal morphology and environments as sea level rose after the ice sheets began to melt. This rise in sea level began about 18,000–20,000 years ago. This chapter will deal with the present situation along the Gulf Coast. Here the "present" is defined as that period from the beginning of historical time up until today. Obviously, there is very little data on the Gulf of Mexico coast from the time when ancient civilizations were flourishing around the Mediterranean Sea. There is, however a great wealth of information on sea-level change over the past century. Change in, and the present position of, sea level is currently a global problem. Thus, the world situation will be briefly considered followed by a detailed look at the situation in the Gulf of Mexico.

Global Situation

During early historical time, advanced civilizations such as the Greeks and Romans were concentrated around the eastern Mediterranean Sea. This area is geologically complex and there is a great deal of tectonic activity. Plate boundary areas exhibit lots of volcanic eruptions and earthquakes. These result in considerable crustal movement and as a result, changes in sea level. There are many locations where structures have been drowned relative to sea level (Fig. 5-1).

Similar factors were in place to cause sea-level change during the time of early civilizations along the west coast of North America. This plate boundary area is also an area of much tectonic activity that causes sea-level changes. A good recent example is the Alaskan earthquake of 1964 (see Fig. 1-1). Keep in mind that tectonic activity may cause rapid and extreme changes in sea level. This fact has been the case throughout geologic time and will continue to be the case in the future. This is a lo-

Figure 5-1. Photo showing old structure in the Venice Lagoon, Italy where sea level has changed markedly since construction. Note that the doors on the structure are about half underwater.

cal phenomenon in sea-level change and is unrelated to global climate change and the cyclicity of glacial activity.

Another very important factor in sea-level rise is the compaction of sediments along the coast. This is basically a regional condition that is focused in the large river deltas of the world. Most of the sediment that accumulates at river mouths is mud that is a combination of silt and clay. The clay particles attract and absorb a considerable amount of water. Because this mud accumulates rapidly it traps a great deal of water, up to 90% by volume. Rapid accumulation of water-containing sediment results in eventual compaction as younger sediments pile onto the older sediment, squeeze the water out, and reduce sediment volume. This reduction in volume causes sea-level rise.

There are deltas in the world where compaction and sea-level rise are not significant because of the amount of sediment supplied equals or even exceeds the rate of compaction. The Amazon Delta is such place. The denudation of the basin, very high rainfall, and cultivation of many new areas has caused sediment supply to the delta to increase over time. This contrasts with many other areas. The Ganges-Brahamaputra Delta along the northeastern coast of the Indian Ocean displays a similar trend although it has many problems of its own.

Two areas where there is considerable compaction and related sea-level rise are the Venice Lagoon in Italy and the outer edge of the Nile Delta in Egypt, although there are differences between the two. The Nile is experiencing a considerable amount of coastal erosion as well as some subsidence due to construction of the Aswan Dam that was designed to provide irrigation water for the Egyptian desert. As a consequence, very little sediment is being delivered to the present Nile Delta. In Venice, the continuing rise in sea level in the Venice lagoon is threatening historical structures. This problem is being caused by a combination of compaction of mud and the elimination of sediment supply due to redirection of two rivers that previously entered the lagoon. In the mid-1950s, Plaza San Marcos in Venice would flood a few times each year due to storm surge in the adjacent Adriatic Sea. Currently, the plaza is flooded nearly every spring tide (Fig. 5-2) which occurs every two weeks. The conditions in Venice have reached the point that extreme measures to ameliorate the problem are underway. Multiple inlets between the barrier islands that form the Venice lagoon are

Figure 5-2. Flooding in Venice in December, 2009. The people are walking on what is a flooded sidewalk. Photograph courtesy Anthony Reisinger.

the avenues where water enters during each tidal cycle and as the result of storm surges in the adjacent open water. The government is in the process of constructing moveable and temporary dams to prevent abnormal increases in water level to keep the city relatively dry. These structures are on a hinge that permits rapid flux of water to lift them and prevent additional water from entering the lagoon. As the current subsides, the hinged dam falls back to the floor of the inlet. These structures are as yet incomplete however, so the efficacy of the approach is currently unknown.

Currently, eustatic or global sea level is rising at an increasing rate (Fig. 5-3). This global rate has shown marked changes during historical time just in North America (Fig. 5-4a). The differences are due to tectonics, rebound, and compaction. Looking at numerous locations on the coast of the Gulf of Mexico (Fig. 5-4b), the general trend is the same but there are some differences in the rates of rise. Overall, sea level has risen globally except where it is overshadowed by tectonics or where glacial rebound is still taking place. There is a considerable amount of variation in the rate of rise but it has been generally slow. The differences are a result of the geological setting of a particular coast. In general, those coasts that received considerable amounts of sediment and that are underlain by unconsolidated sediments experience a rate of sea-level rise that is greater than those coasts that front carbonate platforms or the Precambrian shields that are the ancient core of the continents. Both of these geologic provinces are tectonically stable and are dominated by bedrock that does not experience significant compaction.

Sea-Level Rise Along the Gulf of Mexico Coast

Sea-level changes along the present Gulf of Mexico coast are quite varied, probably more so that any other similarly sized coast in the world. Although there is some tectonic activity, primarily in the form of gravity faults where unstable slopes result in the mass movement of strata, the area is tectonically fairly stable. The changes seen are a combination of natural causes and human interference with natural processes. Sea-level rise has many negative impacts along the Gulf Coast. It drowns wetlands, enables physical erosion (Fig. 5-5), increases salinity in the estuaries, and causes salt water intrusion into important aquifers in addition to other phenomena.

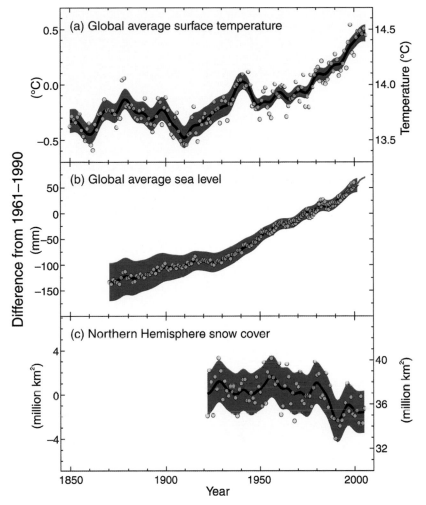

Figure 5-3. Plots of sea level rise, temperature and ice cover in the northern hemisphere over the past century. Ice cover decreases and temperature increases, resulting in a sea level increase. From IPCC (Intergovernmental Panel on Climate Change), 2007. IPCC Fourth Assessment Report: Climate Change 2007, 4 vol. Cambridge, UK: Cambridge University Press.

Early civilizations along the coast of the Gulf of Mexico preceded the completed development of the coastal geomorphology that currently exists. Many Gulf Coast barrier islands had not yet formed. Early Native American populations occupied the coast as much as 12,000 years ago. The Mayans of the Yucatan Peninsula had a sophisticated civilization even before that. Most of these people lived on or very near the shoreline, especially along estuaries where food and transportation were available. Presumably these people experienced sea-level change in the form of a very slow rise. The rise would necessitate changing where they lived as

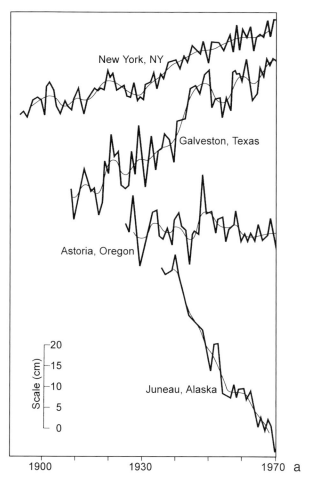

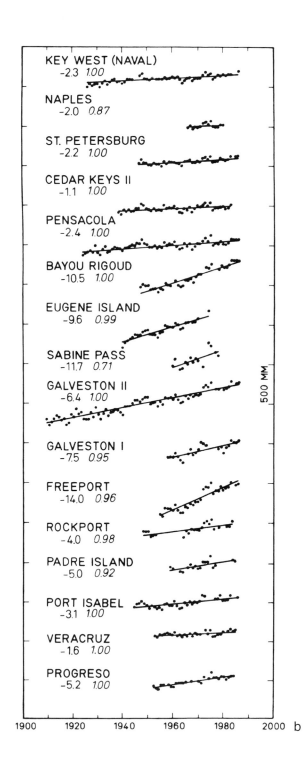

Figure 5-4. (a) Examples of sea level records for various coasts in North America. Compare the Gulf Coast record with those of other coasts. Modified from Hicks, S. D., 1972, On the Classification and Trends of Long Period Sea Level Series. Shore and Beach, 40:20–23; (b) Numerous tide gauge records from around the Gulf of Mexico showing similar trends but different rates. From Emery, K.O. and Aubrey, D.G.,1991, Sea Levels, Land Levels, and Tide Gauges. New York: Springer-Verlag.

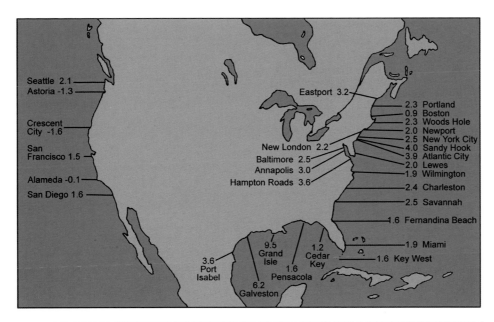

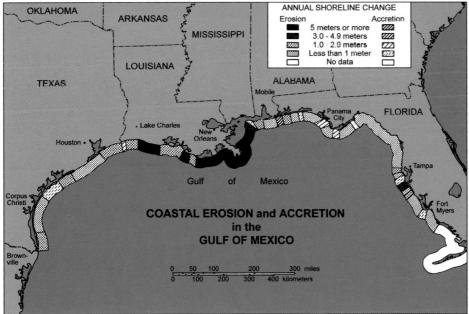

Figure 5-5. (top) Map of U.S. Gulf Coast showing present rates of sea level rise at various locations. From: National Academy of Sciences, 1987, Responding to changes in sea level, Engineering Implications. Washington, D.C.: National Academy of Sciences. (bottom) Map of the U.S. Gulf Coast showing annual rates of shoreline change. Note that there is a parallel with the locations of the rates of sea level rise. From Davis, R. A., 1997, Regional Coastal Morphodynamics Along the United States Gulf of Mexico. Journal of Coastal Research, 13:3595–604.

the centuries passed. This would be in keeping with sea-level curves for the past few thousand years.

The following discussion will consider some of the details of sea-level change on the Gulf Coast since good records have been kept. For practical purposes this does not exceed the past century. Tide gauge records will be used to establish what has taken place and then the reasons for what has happened will be considered. There have been 42 gauges operating along the U.S. Gulf Coast for at least several decades with the oldest going back to nearly 1900 at Galveston, Texas. As in the previous chapters, the coast will be subdivided into geographic and geologic sections that have a significant set of common conditions and circumstances.

Florida Gulf Peninsula

The geologic underpinnings of the Florida Peninsula provide a good explanation for the sea-level changes that have taken place along this coastal region. Recall that this peninsula was once a carbonate platform that was separated from the North American landmass, much like the present Bahama Platform. These carbonate platforms only accumulate limestone and receive no sediment runoff from the nearby continent. In addition, this carbonate platform is not on or near places of tectonic activity in the earth's crust. As a result, the only changes in sea level that would be experienced by this coast would be those that are eustatic (global) in nature. These changes would be the result of global climate change and crustal activity that would change the volume of the oceans.

Tide gauges from this part of the Gulf Coast show the smallest changes in sea level of all of the Gulf of Mexico. Sea level has risen over the period when data were recorded but at rates of only about 2.0–2.2 mm per year (Fig. 5-5); essentially equivalent to the global rate of sea-level rise over the same period. Another advantage of the data from this coast is that most of the tide gauges are based in bedrock so there was no subsidence of the gauges themselves. As a result, this section of the Gulf Coast provides an excellent baseline for comparison of sea-level changes in the other portions of the Gulf.

Northeastern Gulf Coast

The Florida Panhandle and the coasts of Alabama and Mississippi have responded fairly similarly to sea-level change. All of them are essentially wave-dominated, barrier coasts that have significant streams flowing into the Gulf. The two largest streams, the Mobile River and the Appalachicola River, both empty into estuaries; Mobile Bay is the most prominent. These rivers, as well as the other smaller streams, have deposited a considerable amount of sediment along this coast over their history. The consequence of this thick sediment accumulation is that compaction contributes significantly to the present rate of sea-level rise which ranges from 2.4 mm to more than 4.0 mm per year. The rate in this coastal region is lowest on the Florida Panhandle and increases toward the Mississippi Delta.

Mississippi Delta Area

This region has the highest rate of sea-level rise of the entire Gulf of Mexico. There are three primary contributors to the change: 1) compaction of thick mud sequences; 2) withdrawal of fluids; and 3) eustatic sea-level rise. In addition, human activity, mostly associated with the petroleum industry, has exacerbated these changes.

The Mississippi River and its ancestors have been carrying large volumes of sediment to the coast for millions of years. The total accumulation is thousands of meters of mostly mud. This mud is a combination of clay- and silt-sized particles with large amounts of water. As the mud accumulates, its mass causes the underlying fine sediment sequence to compress, driving water from it. As a consequence, the total volume of the deltaic sediments is reduced and the surface is lowered. This lowering of the sediment surface produces a relative rise in sea level.

Since the mid-20th century, there has been extensive and intensive exploration for oil and gas in the sediment sequences of the Mississippi Delta complex. By the turn of the century there were about 40,000 wells, most of which were/are producers. The combined production of these wells removes huge volumes of fluids from the deltaic sediments. The loss of these fluids produces a lowering of the land surface that results in relative sea-level rise. It should be noted that there are now requirements that fluids taken from these sediment sequences must be replaced by surface water from the local area. This will mitigate a portion of the subsidence of the delta region surface.

If subsidence due to compaction and fluid withdrawals is added to the eustatic sea-level rise, the total rate of rise is about 10 mm per year. The vast majority of this rate is due to the regional factors of compaction and fluid withdrawal that contribute almost 8 mm of sea-level rise per year.

The engineering of the Mississippi River for navigation and flood control, and the various construction projects associated with the petroleum industry have produced disastrous circumstances. The multiple dams in the river system are preventing much of the sediment load of from moving down the channel. This greatly reduces the volume of sediment that reaches the delta to nourish the wetlands and help the sediment surface keep up with the rate of subsidence. In addition, levees on the river have been built throughout the delta area. These levees prevent flooding which is good for New Orleans and other areas of where people live in the delta. It is very bad for the wetlands however, because it is the floods that carry sediment across the delta and that will keep up with sea-level rise and keep the wetlands healthy. Any sediment carried to the delta area makes its way to the mouths of the distributaries and then moves out on to the shelf bypassing the wetlands where it will do the most good.

Adding to this problem is the presence of literally thousands of channels that have been dredged for petroleum industry activities. The spoil excavated from these channels is placed along the channel margins producing more levees. Any sediment that is carried along these channels is prevented by these levees from spilling over to nourish wetlands. The net result is that many square kilometers are being lost each year from the Mississippi Delta in particular and the Louisiana coast in general (Fig. 5-6).

Texas Coast

The greatest range in the rate of sea-level rise is found along the Texas coast where it ranges from nearly 12 mm per year at the Texas-Louisiana border to about 4 mm at the Rio Grande area (see Fig. 5-5). There is a distinctly decreasing trend from north to south along the coast. There are multiple reasons for this change.

The East Texas area, including the Houston-Galveston metropolis, is also an area with widespread and high volume fluid withdrawals. Here the fluid withdrawals are due to a combination of oil and gas extraction along with the supply of water to support millions of people. A map of subsidence in this area shows the general pattern associated with the Houston metroplex and localized areas of extreme rates (Fig. 5-7). Places in this highly developed area have subsided such that roads, homes, and infrastructure are now underwater (Fig. 5-8). This is a persistent problem that has not been mitigated significantly.

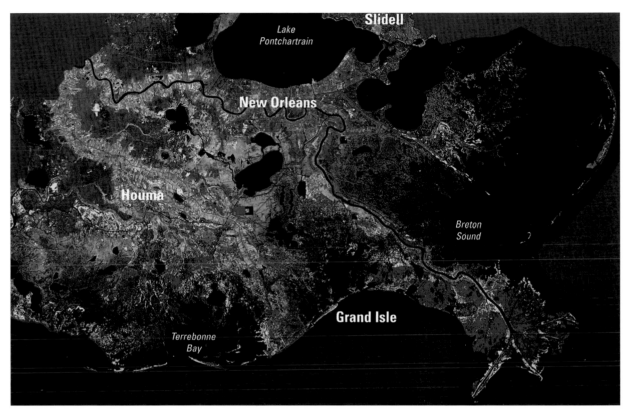

Figure 5-6. Map showing the areas of the Louisiana coast that are being lost due to erosion and sea level rise. The rates increase from the yellow to the red color. Courtesy U.S. Department of Agriculture.

The Texas coastal plain includes many fairly large rivers that have carried abundant sediment to the coast over the past few million years. Although not the volume of the Mississippi Delta area, there is a thick sequence of sediment. Compaction takes place in the same fashion as on the Mississippi but at a slower rate. An additional compaction factor for the greater Houston area is the presence of the city itself. Unlike the Mississippi where only the sediment is causing compaction, there are thousands of huge buildings and tens of thousands of small buildings along with roads, paved parking lots, and other structures in the Houston metroplex. These provide a huge mass that is resting on the thick sediment sequence. The result is compaction, that when combined with fluid withdrawal and eustatic rise in sea level, produces a rapid rate of sea-level rise.

As we proceed southwest along the Texas coast there is a general re-

Figure 5-7. Map of the Houston, Texas area showing the subsidence in this area over the 20th century, much of which is due to withdrawl of water and petroleum. Modified from Harris Galveston Subsidence District, www.hgsubsidence.org/.

duction in the rate of relative sea-level rise but with local exceptions. One such exception is at Baytown, located at the bullseye in Fig. 5.7, where there has been nearly 3 m of subsidence during the 20th century. Much of this is attributed to groundwater removal.

Along the remainder of the Texas coast, rates of sea-level rise are typically 4–5 mm per year which is greater than eustatic sea-level rise. This area has thinner sediment accumulations and only modest petroleum production. These factors have contributed to the greater than

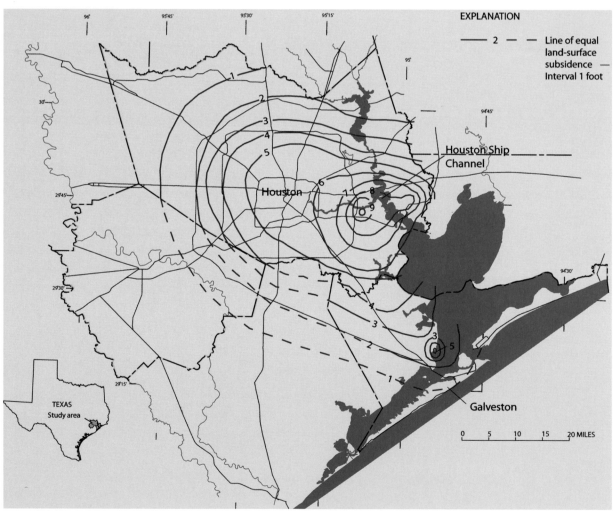

Figure 5-8. A road that now runs into the water. An example of subsidence and sea level rise in the Galveston Bay area of east Texas. Photograph courtesy J. Anderson.

eustatic rate of rise but the rates are also much less than those in East Texas.

Mexico Coast

We have virtually no data on rates of sea-level rise from the coast of Mexico. The rate near the United States border is likely similar to that across the Rio Grande. This is the result of the thick accumulation of sediments in the ancestral river delta(s). Moving to the south, the rate of sea-level rise is influenced by the general lack of large drainage systems emptying into the Gulf and the proximity of volcanic activity. Little deviation from the eustatic rate would be expected except if tectonic activity associated with the volcanic belt was influential.

Sea-level rise in the Yucatan Peninsula is similar to that in the Florida Peninsula in that both were originally isolated carbonate platforms. The thick limestone sequence does not compact and there is no tectonic activity to influence sea-level change. The present rate of rise there is about 2 mm per year, similar to the eustatic rate.

Shoreline Change

Multiple circumstances, in addition to the general rise in sea level, are contributing to shoreline change along the Gulf of Mexico coast (see Fig. 5-5). The following discussion will consider these changes in terms of the general coastal environments in each of the geographic regions. The environments that will be discussed are beaches, wetlands, and tidal flats. Virtually all of the Gulf Coast consists of these environments adjacent to open water.

A major factor in how the shoreline reacts to sea-level rise involves sediment supply. Sediments along this coast come from three main sources: 1) from the mainland via rivers; 2) from reworking of offshore environments in the open Gulf or in the coastal bays; or 3) from the skeletal remains of organisms. All of these make a contribution to shoreline changes on the Gulf Coast.

Florida Gulf Peninsula

Shoreline change along this coast is currently dominated by erosion of open coast beaches. Like most Gulf Coast areas, there are few tidal flats that are significant in width. Estuarine shorelines are dominated by wetlands and wetlands dominate the southern and northern part of this coast. The southwest Florida coast is a tide-dominated mangal (mangrove) coast. It is laced with numerous tidal channels that separate the mangrove communities (Fig. 5-9). These wetlands are stable under present conditions because of the slow rate of sea-level rise and the stability of the carbonate platform. Mangroves are temperature sensitive and do not grow or survive under freezing conditions. Along this coast they only occur northward to about Crystal River, about 80 km north of Tampa.

North of the mangrove area, the wetlands are salt marshes consisting of both *Spartina* and *Juncus*. Just north of Clearwater, the barrier islands and beaches end and the open coast is a *Juncus* high-marsh with numerous tidal creeks and some freshwater discharge in the form of rivers that emerge from springs. These marshes extend for kilometers across from open water to the uplands (Fig. 5-10). Historical aerial photographs show no significant change in this extensive wetland system. The only new sediment being contributed to the system is the organic debris from the marsh plants. As a result, the current slow rate of sea-level rise in this region is not yet causing problems.

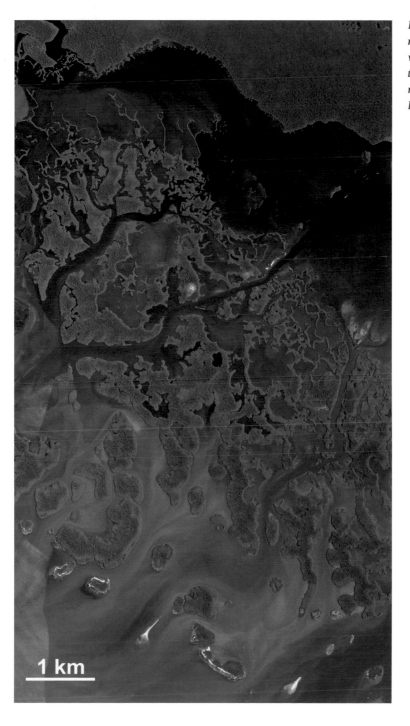

Figure 5-9. Photograph of a mangrove mangal on the south-western Florida coast showing tidal channels that separate the mangrove stands. Courtesy U.S. Department of Agriculture.

Figure 5-10. (a) Panorama aerial photo (infrared) of Juncus marsh on the Big Bend coast of Florida which is a few kilometers wide. (b) Aerial photo of a coastal salt marsh (Spartina alterniflora) in Wacasassa Bay, Florida area showing its extent and the multiple tidal channels. Photo "b" courtesy Cameron Davidson/camerondavidson.com.

a

b

The central portion of this coast is comprised of 30 barrier islands and an equal number of tidal inlets. This is the highly developed tourist and residential part of the Florida Peninsula Gulf Coast. The vast majority of this portion of the coast is eroding and the shoreline is moving landward. There are two primary reasons for this, and neither is sea-level rise. The general absence of sediment on the inner shelf prevents any significant sediment supply to the beaches. Longshore sediment transport along the barrier islands carries abundant sand to the tidal inlets where it is trapped in large sediment bodies. Professor Robert Dean of the University of Florida believes that 80% of the erosion problems of this coast are due to the presence of so many tidal inlets and their efficiency in trapping sediment. Sea-level rise on this coast is not a significant problem. The solution to the beach erosion problems on the Florida Gulf Coast is beach nourishment. Extensive erosion coupled with intensive development has been addressed by placing large amounts of beach-quality sand on the eroding beach (Fig. 5-11). This coast has had more beach nourishment than any other location on the Gulf Coast.

Northeastern Gulf Coast

There has been significant shoreline retreat along much of this coast that is not the result of sea-level rise. This coast has a modest rate of sea-level rise although it is greater than the eustatic rate. It has not, how-

Figure 5-11. Beach nourishment project along the Florida Gulf Coast. The narrow, eroding beach is being replaced by a wide beach by pumping sand from offshore.

ever, caused significant shoreline change. The primary phenomenon that causes erosion here is hurricanes. This area has been a "hurricane alley" over the last part of the 20th century and into the present century. In some places, shoreline retreat has been more than 100 m. At only the present rate of sea-level rise, it would take centuries to cause this amount of shoreline retreat (Fig. 5-12).

This coast is somewhat mixed in its current condition. It includes areas of intense commercial development such as Gulf Shores, Alabama and much of the Florida Panhandle. There are also many kilometers of coastline with only single family residential construction. Some coastal reaches, such as the parks and military properties, have little or no development. Because of the impact of multiple hurricanes, all have experienced significant shoreline retreat. Mitigation has been in the form of beach nourishment, especially on the west end of the Florida Panhandle.

Wetlands are extensive along the backbarrier shoreline and the margins of the estuaries. These wetlands include a combination of *Spartina* and *Juncus* marshes. The rate of sea-level rise has had little influence on these marshes up until now. Hurricanes have also had the most influence on the wetlands. In this case it is the wetlands that are on the landward side of the barrier islands that extend along this entire coast. These storms

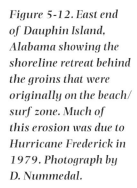

Figure 5-12. East end of Dauphin Island, Alabama showing the shoreline retreat behind the groins that were originally on the beach/ surf zone. Much of this erosion was due to Hurricane Frederick in 1979. Photograph by D. Nummedal.

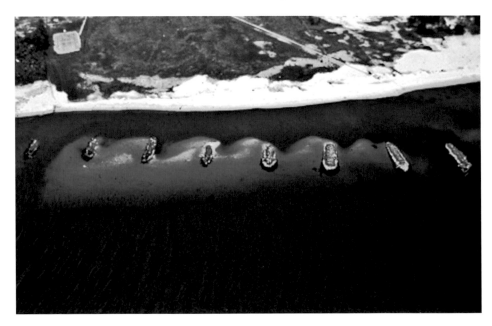

Figure 5-13. Photo showing washover sediment that has extended across the barrier island and on to the wetlands on Dauphin Island, Alabama after Hurricane Frederick. Photograph by D. Nummedal.

have produced extensive washover of sediments that extend all the way into the bays in many locations (Fig. 5-13). This washover sediment has nourished the wetlands and also provided sediment at the proper elevation for salt marsh colonization.

Mississippi Delta Area

This is the most vulnerable portion of the Gulf Coast because the elevation throughout is very low and the rate of sea-level rise is very high. The result is that the State of Louisiana is losing many square kilometers of land every year. Virtually all of the land is lost from the wetlands but a small amount is from the barrier islands.

Erosion of the coast here is essentially a result of the wetlands drowning. Saltmarsh plants can grow only in the upper portion of the tidal range, essentially from neap high tide to spring high tide. Along most of the Gulf Coast the tidal range is less than 1 m and the range of elevation that can support coastal marsh communities is only about 15–20 cm. With the rate of sea-level rise at a centimeter per year it takes only a few years for a marsh to drown.

A river delta is the location of the highest rate of sediment accumulation along any coast. The delivery of sediment to this environment typically provides the appropriate substrate at the appropriate elevation for the development and continuation of salt marsh wetlands. In the case of the Mississippi River there are about 40 dams along its course between the headwaters and the delta. These dams trap much of the sediment that would otherwise make its way to the river mouth and provide for the maintenance of salt marsh communities.

From historical maps it is clear that the rate of growth of the active lobe of the Mississippi Delta has experienced considerable change. As the midwestern part of the United States was cultivated for agriculture during the 19th century, there was a very rapid growth of the active lobe of the delta due to the discharge of large volumes of sediment from the plowed fields. As dams were built in the 20th century for flood control and to aid navigation, there was a marked reduction in the area of this part of the delta (Fig. 5-14) because the dams held so much sediment. This condition continues and is symptomatic of the entire Mississippi Delta and western Louisiana situation. The volume of sediment now discharged at the mouth of the Mississippi is about half of what it was before the entire system was engineered. The bottom line is that the marshes are drowning because there is not enough sediment being emptied in to the delta region.

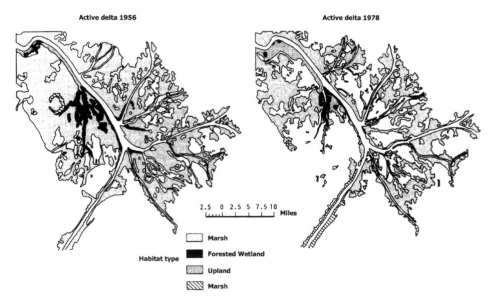

Figure 5-14. Maps showing the areal reduction of the active lobe of the Mississippi Delta from 1956 to 1978. Courtesy U.S. Fish and Wildlife Service, National Coastal Ecosystem Team.

Figure 5-15. Photo of offshore structures that were installed to mitigate the erosion problem in western Louisiana. Photograph courtesy G. Stone.

While the loss of wetlands is the most critical problem in the Mississippi Delta region, it is not the only problem. The wetlands are somewhat protected by barrier island systems. These barriers are formed by the reworking of older delta lobes that preceded the present active lobe. Most of these barriers are west of the present active lobe and another smaller barrier is north of the modern lobe. These barriers protect the wetlands from storms including to some extent hurricanes. Unfortunately, the barrier islands are also quite vulnerable to erosion and sea-level rise. The relatively large barriers have communities and petroleum installations on them. They are low in elevation so even a modest storm surge can produce washovers and beach erosion. Recently, efforts have been made to try to stabilize these islands by a combination of beach nourishment and protective structures (Fig. 5-15). The results have not been good. The smaller chain, the Chandeleur Islands, has been reworked from the St. Bernard lobe that accumulated about a thousand years ago. These barriers are completely natural and low-lying with numerous tidal channels between them (Fig. 5-16). They are washed over regularly, even by modest storms. As time passes these barriers are being destroyed. These examples illus-

Figure 5-16. Photo of Chandeleur Islands on the Mississippi Delta showing extensive overwash and numerous channels. Photograph courtesy G. Stone.

trate that the Louisiana coast is in great jeopardy. It is very vulnerable, very fragile, and is being lost at a tremendous rate.

Texas Coast

There is great range in the response of this area to the present sea-level conditions. For purposes of discussion it is best to divide this region into the eastern part—East Texas and westernmost Louisiana—and a southern part which is the rest of the Texas coast. The reason for this separation is that sea-level rise is much more rapid in the eastern part and the response is dramatic. To the south, the rate of sea-level rise is only a bit higher than the eustatic rate and the response is modest.

Sea-level rise in western Louisiana and East Texas to about Baytown is about 3–4 times the global rate. This is due to a combination of compaction of fairly muddy sediment carried to this coast from the Mississippi Delta, withdrawal of fluids, both petroleum and groundwater, and eustatic rise. The open coast in this area experienced a considerable amount of erosion during the late 20th century that continues today. Holly Beach in western Louisiana is a developed area that has been nourished to mitigate erosion however, Hurricane Rita destroyed this beach. The Bolivar Peninsula in eastern Texas has also experienced high erosion rates over the same time frame. The coastal highway (TX 287) has been essentially destroyed and has been closed for many years. Hurricane Ike in 2008 added to this problem.

The area near the city of Galveston and Galveston Bay has also experienced a great deal of erosion and subsidence. After the hurricane of 1900 essentially destroyed Galveston, two measures were taken to restore and protect the city. The entire community was raised almost 2 m (Fig. 5-17) in an effort to prevent future hurricane inundation of the city. In addition, a seawall was constructed for additional protection. This structure has lasted more than 100 years and is considered by many as an engineering marvel (Fig. 5-18). Even though the city has been protected for some time, the west end of Galveston Island and beyond is eroding and the area has been overtopped by multiple hurricanes. It is difficult to assess the individual roles of hurricanes and other severe storms in this deterioration of the coast as compared to the relatively high rate of sea-level rise.

The bays and wetlands of this coastal region have experienced considerable deterioration during the late 20th and early 21st centuries.

Figure 5-17. Photo showing the different elevations of the city of Galveston as reconstruction was taking place after the hurricane of 1900. The elevation of the city was raised about 2 meters to protect it from future storm surge. Photograph courtesy U.S. Army, Corps of Engineers.

Figure 5-18. Galveston sea wall constructed after the hurricane of 1900. It is considered as an engineering marvel and has withstood the test of time but was washed over by Hurricane Ike in 2008.

Wetlands have been drowned by rapid sea-level rise in the absence of sediment to nourish the marshes. Subsidence in the Galveston Bay area has been so great that some roads and homes have become inundated (see for example Fig. 1-4). There have been attempts to mitigate wetland loss through construction of salt marshes. This effort has centered on planting appropriate species on substrate that has been molded to the desired elevations and substrate type (Fig. 5-19). This new wetland appears to be doing well but its lifetime will be finite. As sea level rises in the Galveston Bay area there is no source of sediments away from the delta of the Trinity River. The result is that these constructed marshes will drown in the same fashion as those on the present Mississippi Delta.

Another coastal reach in this area that has experienced extreme erosion rates is at Sargent Beach located about 100 km west of Galveston. Here a coastal community was developed in the 1950s and 1960s. As time passed, erosion moved the shoreline on the open Gulf hundreds of meters landward and at least two blocks of houses were destroyed (Fig. 5-20). This was partly due to compaction because much of the development was on muddy sediment deposited on old wetlands, and there was a lack of new sediment. The rate of shoreline migration was such that there was concern about it impacting the Intracoastal Waterway (ICW). The solution developed by the U.S. Army Corps of Engineers was to construct a protective wall along the impacted area. This project was completed in 1998 (Fig. 5-21) and stopped shoreline retreat along this part of the Texas coast.

In general, the central and southern coast of Texas is eroding but is much more stable than the East Texas coast. The rate of sea-level rise is higher than the eustatic rates. Sediment accumulation in this area throughout the Quaternary Period was dominated by fluvial-deltaic environments. These sediments were delivered by the numerous rivers that crossed the coastal plain and were generally mud with some sand.

Figure 5-19. Photo of constructed coastal marsh in the Galveston Bay area produced by HDR Engineering. These projects help to mitigate destruction of coastal wetlands. Photograph courtesy Lanmon Aerial Photography.

Figure 5-20. Extensive erosion at Sargent Beach, Texas: (top) in close up view and (bottom) an aerial view across the area. The muddy shore-line represents older marsh deposits that were once behind the barrier sands. Many houses have been removed by erosion since construction began in the late 1950s.

These sequences have compacted over the past several hundred thousand years. In addition, there has been fluid withdrawal by the petroleum industry but at rates much below those of East Texas and the Mississippi Delta area.

At the present time, beaches are eroding and sediment is being lost through both offshore transport and washover. There are bayhead deltas in most of the estuaries that carry significant sediment to the estuaries.

The amount has been reduced by dams along the rivers such as the Nueces, Colorado, Guadalupe, and others. Nevertheless, the wetlands associated with these bayhead deltas are reasonably healthy. The washover sediment also supports wetlands on the landward side of the barrier islands. Black mangroves are scattered along this part of the coast in addition to the usual salt marshes.

It should be noted that climate change in South Texas has taken place during the past several thousand years. This is now a semi-arid area and river systems have essentially dried up. A good example is the present Baffin Bay. Its geomorphology shows that at one time it was an estuary fed by multiple streams (Fig. 5-22). At present these are only intermittent streams and Baffin Bay is hypersaline, not brackish, as is typical of estuaries.

Figure 5-21. The sea wall built to stop erosion at Sargent Beach, Texas just after completion. In only a few years it was completely covered by sand. Photograph by Stephan Meyers courtesy Texas Sea Grant Program.

Figure 5-22. Aerial photograph of Baffin Bay showing multiple relict channels and bays that were fed by streams that are now only intermittent. This is a good example of the influence of climate change on geomorphic change. Courtesy U.S. Department of Agriculture.

Mexico Coast

The coast of Mexico is generally more stable than that of the United States. South of the Rio Grande, the open coast beaches are eroding slightly but not as rapidly as those to the north. The Tabasco coast is still prograding due to sediment supply coming from the adjacent rivers. From this area across the Yucatan Peninsula there are barrier islands. Those on the Yucatan are pure carbonate. Except for hurricanes, the rate of shoreline retreat is slow.

Much of the Mexico coast has wetlands dominated by mangroves. Because these species are resistant to erosion and are tolerant of modest sealevel change, the wetlands here are stable and expanding in some locations. Unlike the Texas coast, there are multiple mangrove taxa including black (*Avicennia germinans*) and red (*Rhizophora mangle*).

To summarize, the present situation across the Gulf Coast shows a wide range in conditions. Some areas are essentially stable and reflect only the eustatic rate of sea-level rise. By contrast, there are areas where compaction and fluid withdrawal are important factors and the rate is up to five times the global rate. East Texas and Louisiana are the most affected coasts and the Florida and Yucatan peninsulas are the least.

Addressing the Problems

The problems along the Gulf of Mexico coast are obviously numerous. The attempts at addressing these problems are also numerous. Most of the problematic situations can be categorized into one of two different processes: 1) erosion of beaches on the open coast; and 2) loss of wetlands. Both of these problems are at least partly due to sea-level rise.

Beach Erosion

Beaches are one of the most important tourist destinations throughout the world. This is especially true on the Gulf Coast. The Florida Gulf Coast, the Mississippi and Alabama coasts, and the Texas coast are all experiencing beach erosion. Because the economics of coastal tourism relies on the presence of beaches, this problem must be mitigated. It should be noted that not only are the beaches eroding, but also many residences and commercial buildings are being damaged or destroyed. Granted, tropical storms and hurricanes are the apparent major factor in this erosion but it is being facilitated by the slow and constant rise of sea-level. This is most apparent in East Texas where the rate of rise is 8–12 mm/yr, several times the average rate in the Gulf.

Mitigating this problem is difficult, costly, and temporary. In the past, the standard approach was building structures such as seawalls, groins, rip-rap, or some combination (Fig. 5-23). These structures are costly, unsightly, and do not really provide for maintenance and/or protection of the beach, only the temporary protection of the buildings. The current approach to this erosion problem is to nourish the beaches by placing

Figure 5-23. Photos showing (top) a seawall and (bottom) groins that were constructed to help protect and preserve the beach environment. Generally, the results of these types of structures were not positive. That is a main reason why coastal protection is now dominated by beach nourishment.

large volumes of beach-quality sand on the beach, essentially building a new beach (Fig. 5-24). These beaches provide area for recreation and also protect the upland environment. The downside of this approach to the erosion problem is the combination of the coast and the durability of a nourished beach. A nourished beach can cost millions of dollars and only

Figure 5-24. A nourished beach along the Florida peninsula. Such projects have been generally successful along this coast but they are expensive and are temporary.

last a few years. It is common for this type of project to cover one to several kilometers of shoreline. The total amount of sand required for such a project ranges from hundreds of thousands to more than a million cubic meters. The cost per cubic meter can be up to $25 depending on the method of delivery and the distance of the sand source from the construction site. These projects typically last only a few years before re-nourishment is necessary. Severe storms are a major factor in their performance. Many people question their value but studies have shown that beach nourishment is cost-effective.

Wetland Loss

The biggest coastal problem associated with sea-level rise is the loss of wetlands (see Fig. 5-6). On a global scale, wetland area will probably be reasonably stable until 2025. Like beaches, wetlands are eroded by storm waves but their widespread loss is primarily a sea-level problem. Although mangroves are a prominent wetland on the southern coast of Florida and in much of Mexico, these large and sturdy trees can withstand both storms and sea-level rise rather well. This discussion will disregard mangrove erosion as a serious problem.

As mentioned earlier, coastal marsh vegetation has a very narrow window of habitat. It colonizes only the upper part of the intertidal environment. On most of the Gulf Coast this is only about 15–20 cm in elevation. In places where sea level is rising a centimeter per year, this means that the marsh must also accumulate sediment at the same rate to sustain the environment. The dams on the Mississippi River and the numerous levees throughout the Mississippi Delta drastically reduce the sediment delivered to the marshes.

A valuable concept that was originally applied to sea-level rise over coral reefs by A. C. Neumann of the University of North Carolina can also be used to describe the fate of salt marshes. This is the principle of "catch up, keep up, or give up." That is, if sea level rises at 5 mm/yr and the marsh has only been accumulating sediment at the rate of 3 mm/yr but then increases to 5mm/yr, it will "catch up" and be a viable marsh community. If the marsh accumulates sediment at the same rate as sea-level rise, it will "keep up" and also be viable. On the other hand, if the rate of sea-level rise is greater than the rate of sediment accumulation on the marsh substrate, the marsh will drown and that community will be lost; the "give up" situation (Fig. 5-25). This is what is happening in most of the wetlands on the Louisiana and East Texas coastal zones. Louisiana is presently losing 50–100 km^2 per year. Saltmarsh coasts where sea-level rise is at or near the eustatic rate have a few decades before the "give up" situation presents itself.

Saltmarsh loss is mitigated by constructing new wetlands. This is also expensive and cannot sustain the environment in most locations. Saltmarsh plants have been planted along portions of Tampa Bay in Florida and mangroves have been planted in places on the Mississippi Delta where marshes cannot persist. Probably Galveston Bay in Texas has been the location of the most widespread and successful constructed marshes. In fact, hundreds of acres of marsh have been constructed in this estuary. The procedure is to place sediment in low mounds with their elevations cresting in the upper intertidal zone. The plants are then planted in the appropriate area. This provides a marsh environment that is comprised of many circular segments (Fig. 5-26). Shallow, subtidal channels separate the marsh per the requirements of the U.S. Fish and Wildlife Service. The rationale is that this format will enhance the fishery in the estuaries by providing a maximum area where food will be present (R. Thomas,

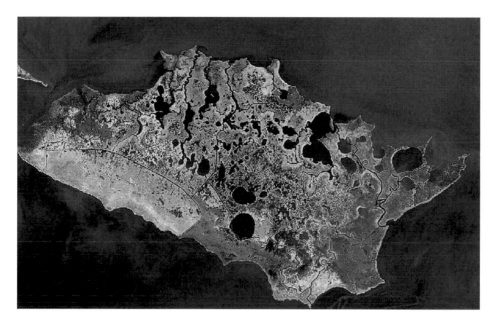

Figure 5-25. (top) High altitude image of drowning marsh. Courtesy U.S. National Wetlands Research Center and Louisiana Department of Environmental Quality; and (bottom) a portion of the drowning of the marsh coast in the Mississippi Delta area. The lack of definition on the margins of the marsh is an indication that they are drowning.

Figure 5-26. (top) Initial stages of marsh construction in Galveston Bay and (bottom) well-established marsh community that resulted after less than two years. Photographs courtesy Lanmon Aerial Photography.

personal communication). The problem with this approach is that there is no natural sediment supply for these new marshes so the high rate of sea-level rise in this area will drown the marsh in 20 years or less. The bottom line is that while we can mitigate the coastal problems with huge expenditures, these approaches are short-lived.

Summary

The highest rate of sea-level rise since the planet became extensively populated is occurring now. There are locations where the rate of inundation of coastal environments is slow and others where it is high and very destructive. The combinations of sea-level rise with fluid withdrawal and sediment compaction lead to regional differences in the rate of sea-level rise. The Florida Peninsula is experiencing the slowest rise and the Louisiana coast is undergoing the highest. Areas of eastern Texas are also in jeopardy.

Various environments react differently to sea-level rise. Coastal wetlands are the most vulnerable because their vegetation is very much controlled by sea level. Lack of sediment availability is causing these wetlands to drown. In some places this is occurring so rapidly that the marsh cannot respond and migrate with the rise in sea-level. Estuaries are becoming more saline which is causing unbalance in the biota that lives in this environment. This is also increasing the effects and rate of saltwater intrusion along the coast.

Rising sea level also has an effect on human activities such as beach nourishment, location and construction of coastal structures, and shipping. The effect of all of these changes is very expensive and is very destructive to the natural environment.

Important Readings

Bird, E. C. F. 1993. *Submerging Coasts; the Effects of a Rising Sea Level on Coastal Environments.* Chichester, UK: John Wiley. A global look at coasts that are in trouble because of sea-level rise.

Boesch, D. F., M. N. Josselyn, A. J. Mehta, J. T. Morris, W. K. Nuttle, C. A. Cimestad, and D. J. P. Swift. 1994. *Scientific Assessment of Coastal Wetland Loss, Restoration and Management in Louisiana.* Journal of Coastal Research Special Issue No. 20. West Palm Beach, Fla.: Coastal Education and Research Foundation. This volume addresses the problems of the most critically eroding part of the Gulf of Mexico coast.

Douglas, B. C., M. S. Kearney, and S. P. Leatherman, (eds.). 2001. *Sea Level Rise: History and Consequences.* International Geophysics Series No. 75. New York: Academic Press. A series of topical papers that consider various factors in sea-level rise and what the future holds for each.

Emery, K. O. and D. G. Aubrey. 1991. *Sea Levels and Tide Gauges.* New York: Springer-Verlag. Detailed information on the measured sea-level data. Numerous global examples with a large section on the Gulf of Mexico.

IPCC (Intergovernmental Panel on Climate Change) Working Group 1. 2001. *Climate Change 2001: The Scientific Basis.* New York: Cambridge University Press. A landmark volume that stresses climate change but also includes sections on sea-level rise problems.

Knapp, B. and M. Dunne. 2005. *America's Wetland: Louisiana's Vanishing Coast.* Baton Rouge: Louisiana State University Press. A beautifully illustrated volume that chronicles all of the problems with the present situation on this most vulnerable coast.

Moulton, D. W., T. E.,Dahl, and T. M. Dall, 1997. *Texas Coastal Wetlands: Status and Trends, Mid-1950s to Early 1990s.* Washington, D.C.: U.S. Department of Interior, U.S. Fish and Wildlife Service. Extensive look at the historical changes and current situation in Texas coastal wetlands.

Perillo, G. M. E., Wolanski, E., Cahoon, D. R., and Brinson, M. M. (eds.). 2009. *Coastal Wetlands: An Integrated Ecosystem Approach.* Elsevier Science, Amsterdam. This is the most current and comprehensive volume on coastal wetlands with the chapters written by global experts.

Pilkey, O. H. (ed.). 1983–91. *Living with the* [State] *Shore.* Durham, N.C.: Duke University Press. A series of books on current conditions along the coasts of individual states.

6

What Is Next?

This chapter deals with the future. Three aspects of what will take place along the Gulf of Mexico coast must be considered. This discussion will be aimed at the turn of the next century, the typical time frame for most of the published predictions. First and foremost, what will be the position of sea level at the year 2100? Then, what is the impact of that sea-level position will have on coastal environments, both natural and built, must be considered. The final consideration will be how society will cope with this cause and effect situation. As is typically the case, cost will be a major factor in the extent and nature of the mitigation that is proposed.

Predictions are always difficult. As an example, a prominent expert on sea-level rise stated that by the year 2075 a 100-year storm of medium intensity would overtop the Galveston sea wall. Hurricane Ike, a category 3 hurricane, did the job in 2008. Those predictions that are of a global nature are especially difficult. In the case of sea level, there are numerous variables that contribute to its change. The common feeling is that climate change and the formation or reduction of ice sheets is what controls sea-level change over millennia. As explained in Chapter 1, there are many factors that contribute to the position of sea level. Changes in ice sheets result in some of the most rapid changes that contribute significantly to global sea-level change. This is the situation as far as global change is concerned. Historical records show that there has been an increase in the rate of sea-level rise through most of the 20th century (see Fig. 5-3), and especially during the last decade.

In point of fact, although the eustatic rate of rise is increasing and will be a long-term problem for many coasts if the trend continues, it is not the major problem for the Gulf of Mexico. The biggest problem here is subsidence in the area of the Mississippi Delta region. This part of the coast has the serious combination of problems in compaction and fluid withdrawal added to the global situation of melting ice sheets and oceanic warming.

There are other coastal reaches of the Gulf of Mexico that have longer term problems, especially with the beaches and wetlands.

There are data to indicate that changes in the composition of the atmosphere and the relationships among celestial bodies are significant contributors to global warming and as a result, to sea-level rise. All of the responses to sea-level rise are essentially negative in their impacts on the natural environment and on the economy of the coast. Therefore, it is important to make every effort to accurately predict the future of sea-level rise (Fig. 6-1). Other detrimental contributors can be indirect results of sea-level rise and are exacerbated by human activities.

To this end, global sea-level rise predictions will only provide an indication of some of the potential problems along the Gulf Coast. As a result, predicting what will happen along this diverse coast is compounded by local and regional variability. This discussion will follow the format of previous chapters and discuss the future by region.

Figure 6-1. Flow chart showing major factors that lead to sea level rise and its consequences.

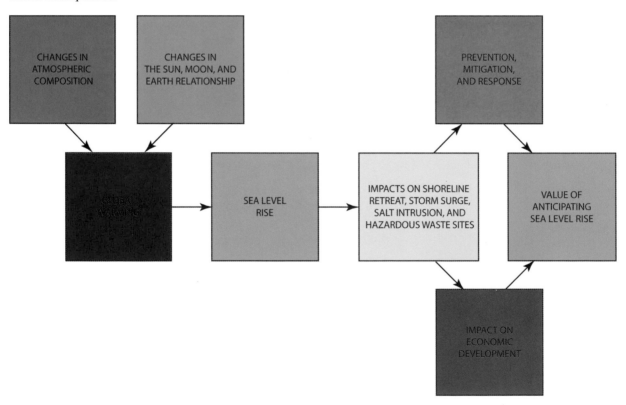

Future Sea-Level Rise

Various governmental organizations and high-level international committees have been concerned with trying to predict the rate of sea-level rise. Most of this effort began in the 1980s when it was recognized that the rate of sea-level rise was increasing. Data showed that the mean annual temperature of the globe was increasing, the temperature of the oceans was increasing, and the areas of ice sheets, glaciers, and sea ice were being reduced. All of these factors contribute to a rise in eustatic sea level. The lead domestic organizations involved in the prediction of the rise are the U.S. Environmental Protection Agency (EPA) and the National Research Council (NRC). Globally, the Intergovernmental Panel on Climate Change (IPCC) is the leading group working on this problem.

An early conference of representatives from major coastal cities convened in Venice, Italy (an obvious choice of locations) in 1990 to address the problem. As is the case for most such conferences or committee studies, the main time frame was the 21st century. Their forecasts have been promulgated widely. Because the effort was directed to the global crisis, the emphasis was on the perceived rapid factors, glacial melting and increasing ocean temperature (Fig. 6-2). The Antarctic ice sheets were predicted to be the biggest contributor followed closely by other ice sheets, except for Greenland which was considered separately. According to some experts, melting of the Greenland ice sheets may be the largest contributor at the present time. Thermal expansion of the oceans due to increased global temperature is the other major contributor. The range of each envelope of change is large because of the uncertainty of future climate changes. This type of uncertainty is present in all of the predictions about sea-level change and there is really nothing that can change it (Fig. 6-3).

The IPCC has produced two very important reports over the early 21st century: IPCC 2001 and IPCC 2007. These are considered to represent the best that the world scientific community has to offer in terms of predictions of what will happen over the rest of the century as a result of sea-level change. Data show that the average global rate of rise over the 20th century was 1.7 mm/yr. Sea level was rising at about twice that rate during the 1990s (Fig. 6-4), an observation that some investigators interpret as an anomaly.

Figure 6-2. Graphs that show the predicted envelopes for sea level rise over the 21ˢᵗ century based on various glacial contributions and oceanic temperature. Modified from Ince, M. (ed.), 1990, The Rising Seas. Proc. Of Cities on Water Conference, Venice, Italy, London: Earthscan Publications.

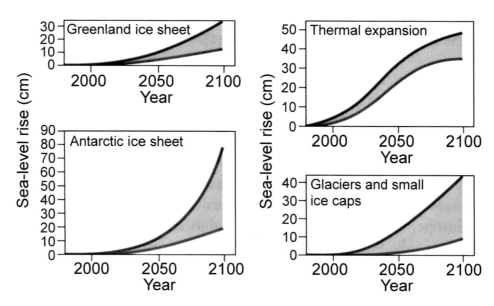

Figure 6-3. Predictions of sea level rise over the 21ˢᵗ century as expressed by the Intergovernmental Panel on Climatic Change IPCC 2007. IPCC Fourth Assessment Report: Climate Change 2007, 4 vol. Cambridge, UK: Cambridge University Press.

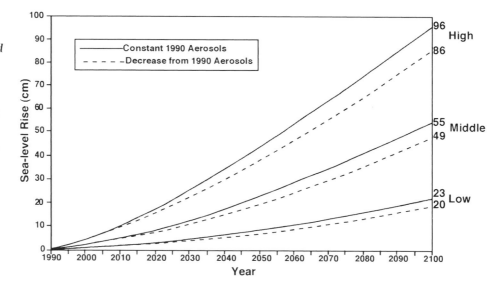

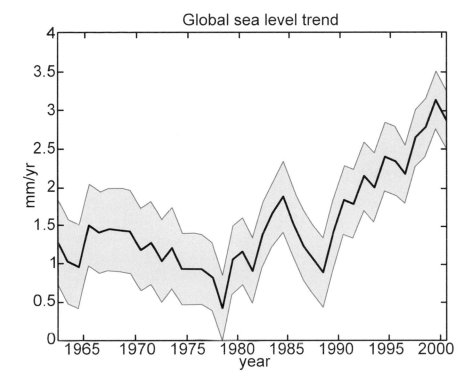

Figure 6-4. Global sea level for the last part of the 20th century showing the rapid increase at the end. From Merrifield, M.A., S. T. Merrifield, and G. T. Mitchum, 2009, An anomalous recent acceleration of global sea level rise. Journal of Climate, 22:5772–781. © American Meteorological Society. Reprinted with permission.

For the purposes of discussion two different sea-level rise scenarios for the current century will be considered: a rise of 0.5 m and a rise of 1.0 m. These are the intermediate and high predictions to the year 2100 as projected by IPCC 2001. Although these predicted rates are higher than those presented in the 2007 IPCC report, they are probably about what can be expected for the Gulf of Mexico. In some locations the rate might be higher, such as East Texas and Louisiana and in others, such as the coast of the Florida Peninsula, the rate will be less. The open coast situation will be discussed separately from the estuaries, lagoons, and bays that typically respond differently.

A study by the U. S. Geological Survey has assessed the vulnerability of the U.S. Gulf of Mexico Coast to sea-level rise over this century. This study used six variables to determine vulnerability: 1) geomorphology; 2) coastal slope; 3) relative sea-level change; 4) shoreline change; 5) mean tidal range; and 6) mean wave height. The result was that 42% of the Gulf

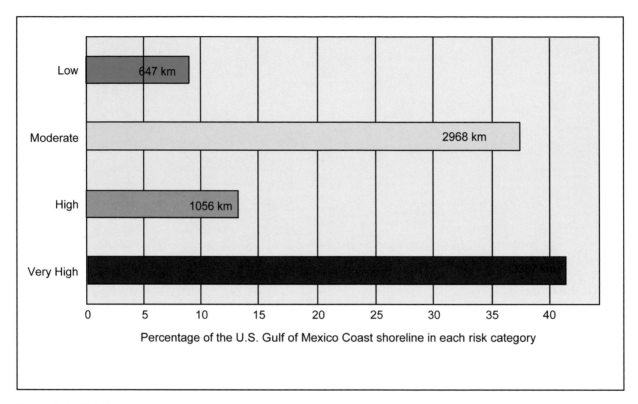

Figure 6-5. Distribution of vulnerability categories as determined by the U.S.G.S., Coastal Research Team, Woods Hole Center. From: Theiler, E. F. and E. S. Hammer-Klose. 2000. National Assessment of Coastal Vulnerability to Sea-Level Rise, Preliminary Results for the U.S. Gulf of Mexico Coast. USGS Open File Report 00–129. Washington, D.C.: U.S. Department of Interior, U.S. Geological Survey.

Coast is at very high risk and 13% is at high risk to sea-level rise. Only 8% is at low risk (Fig. 6-5). As expected, the Louisiana and East Texas coasts have the highest risk and the Florida peninsular coast is the lowest.

Florida Gulf Peninsula

This portion of the Gulf Coast can be divided into three distinctly different sections: 1) the southwest coast which is tide-dominated and contains numerous mangrove islands separated by tidal channels on the open coast; 2) the developed coast which is dominated by barrier island and associated tidal inlets; and 3) the northern Big Bend coast which is an open-coast *Juncus* marsh. Unlike the rest of the U.S. Gulf Coast, there is no local sea-level change along the Florida Peninsula because it is a stable carbonate platform. Only the eustatic sea-level change will have an effect. The total rise over the 21st century will be the lowest of the U.S. Gulf Coast.

Except for Tampa Bay and Charlotte Harbor, there is no significant freshwater input to this coast. Sea-level rise will cause salinity to increase in each of these estuaries and will influence the organisms living there. Brackish water fauna such as oysters will be more limited in their distribution. This coast currently has problems with saltwater intrusion into the very permeable limestone of the Florida Platform. The increase in sea level, regardless of the rate or amount, will increase this intrusion into the freshwater aquifer and cause water supply problems for the highly populated coastal areas.

Mangrove Coast

The least affected of these three coastal sections under either of the sea-level rise scenarios is the mangrove coast south of Marco Island. Mangroves have extensive and deep root systems and they can tolerate sea-level change. The most likely problem along this coast is the rate of sediment accumulation at the base of the mangrove communities (mangals). There is little sediment runoff so most of the new accumulation depends on shell production, primarily oysters and barnacles, and the leaf litter from the mangroves themselves. Peat is common in this area and provides a firm protection from increasing wave action as sea level rises. The high sea-level rise scenario will be potentially more destructive to the mangroves than the moderate scenario because vulnerability to wave action will increase. However, this part of the Gulf Coast has a very low-energy wave climate.

Barrier-Inlet Coast

The barrier-inlet coast of Florida will experience a range of responses to the forecasted sea-level rise. The open coast is dominated by beaches that typically front extensively developed barrier islands. This is the part of the Florida Gulf Coast that attracts millions of people each year for recreational purposes. It is a major part of the Florida economy. It is also a part of the Florida coast where sediment availability is limited. This dearth of sediment presents problems for beach nourishment projects which are currently a widespread solution for beach erosion. This approach to maintaining beaches is very expensive and obviously dependent on sand availability.

Beach erosion along this coast will certainly be accelerated and the higher sea-level rise scenario will be much worse than the moderate scenario. Even though wave energy along this coast is low, increasing sea level permits this wave energy to increasingly influence the sand beaches and adjacent small dunes. It is unlikely that beach nourishment can continue to be the answer to the beach erosion problem due to the combination of cost and sediment availability. In a few decades, certainly by 2050, there will probably have to be a new approach to stabilizing this coast. The only real solution, other than abandoning the coast, is construction of barriers to protect landward buildings and infrastructure. This was the approach prior to the advent of beach nourishment projects. Reverting back to this type of protection would be aesthetically unpleasing and would be expensive. The retreat-from-the-coast option is unlikely because of the investment in the built environment and the need for attracting tourists to the coast. Although both sea-level rise scenarios will eventually result in the need for protection, the moderate rate of rise will prolong the current situation. However, even the moderate rate of rise will eventually result in the need for different approaches to maintaining the beaches.

The bays on the landward edge of the barrier islands and mainland represent a much different scenario for future sea-level positions. These shallow bays are fed by small rivers or in some cases there is no freshwater input. The coastal wetlands are a combination of mangrove mangals in the more southerly areas and salt marshes throughout. The salt marshes are vegetated by a combination of *Spartina* and *Juncus.*

As in the more southern part of this coast, mangroves will be only moderately affected by either sea-level rise scenario. If the rate of rise is fairly uniform over the century, there will be opportunity for the mangroves to encroach upon the slowly drowning mainland shoreline. The salt marsh will be more vulnerable to drowning due to the absence of sediment coming into the bays. There is considerable development along much of the coast of both the mainland and the backbarrier (Fig. 6-6) that prevents landward migration of either mangroves or saltmarsh vegetation. The result will be a major reduction of wetland environments regardless of the rate at which sea level rises. There is little that can be done or that will be done to alleviate this condition. Wetlands will be

Figure 6-6. Intensive and extensive development along the Gulf Coast of central Florida. This dredge and fill development is no longer permitted.

significantly reduced along this coast as sea level rises during the 21st century.

Big Bend Coast

The open coast salt marsh that characterizes the Big Bend coast of Florida is extremely vulnerable to sea-level rise. Mangroves are not tolerant of freezing and are not present on this part of the Florida coast. This area is remote and essentially without development. The marsh is several kilometers wide, which means that the surface elevation is flat with no significant slope. The marsh is dominated by the high marsh plant *Juncus*. There is virtually no mineral sediment delivered to the marsh from either the Gulf or rivers. The adjacent inner shelf on the Gulf is sediment-starved so storms cannot deliver significant amounts of sediment. The rivers are short except for the Suwannee River and none of them carry much sediment.

The result is that the rates predicted in either the moderate or high sea-level rise scenarios will cause much of the marsh environment to drown over this century. There is really no way to mitigate this condition. It is possible that a marsh fringe might encroach along the present mainland because it is very low in elevation but it will likely be only modest in extent.

Northeastern Gulf Coast

The northeastern Gulf Coast includes the Florida Panhandle and the Alabama and Mississippi coasts. It is a barrier island coast with multiple estuaries, the largest of which is Mobile Bay. The two largest rivers that deliver substantial sediment to the coast are the Appalachicola and Mobile. The projected sea-level rise over this century will have a substantial impact on these coastal environments.

Like the Florida Peninsula, barrier beaches here are important tourist destinations, except on the Mississippi coast where the gambling industry is the main attraction for tourists. Erosion has been a problem (see Fig. 5-5) and will be exacerbated by the sea-level rise. Beach nourishment has been moderately successful with relatively abundant nearshore sand sources. Because the inner shelf here is steeper than anywhere else along the Gulf Coast, increasing sea-level rise will have a major effect on beach erosion by increasing wave energy. Beach nourishment will not be as effective for mitigating erosion in the future.

The coast of the bays and estuaries in this area is dominated by the salt marsh environment. Both sea-level scenarios will drown most of this saltmarsh system because of a lack of sediment. Exceptions will be in the areas of the Appalachicola and Mobile river deltas (Fig. 6-7). In these areas marsh vegetation should be able to maintain itself, especially in the Mobile Bay area because the Mobile River carries more sediment than the Appachicola River. Salinity in Mobile Bay will increase with an accompanying change in the estuarine community.

There will be some variation in the rate of relative sea-level rise on this coastal region because compaction of sediment will produce subsidence. This process will produce modest changes that will be most prominent on the Alabama coast over the paleochannel of the ancestral Mobile River.

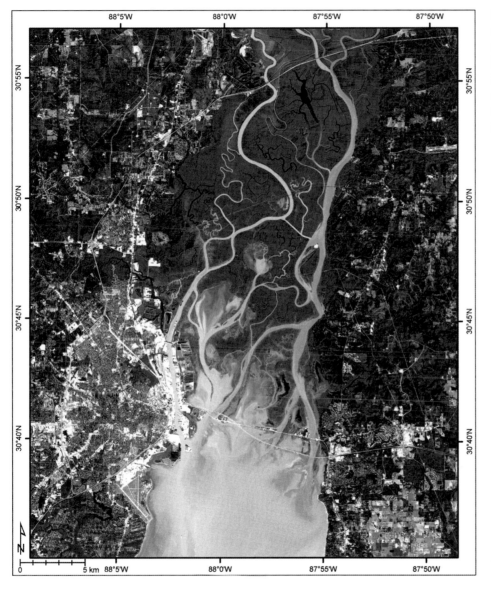

Figure 6-7. Extensive marsh associated with the Mobile River delta in Mobile Bay, Alabama. Courtesy Alan Criss, Gulf Coast Geospatial Center, University of Southern Mississippi.

Mississippi Delta Area

Without question, the response to sea-level rise in the Mississippi Delta will be the most complicated and potentially the most devastating on the entire Gulf of Mexico coast. Complicating the problem is the varied ways in which human activities have negatively impacted the delta region. In this region, sea-level rise is currently 10 mm/yr, more than four times the global average, with local rates up to 20 mm/yr. Compounding this extreme rate of relative sea-level rise are the many changes in the natural environment that have been caused by activities to foster and maintain navigation in the Mississippi River system and those associated with oil and gas exploration and production.

This region includes the entire Louisiana coast, both the Mississippi Delta area and the Chenier Plain to the west. Both owe much of their character and development to the Mississippi River. Most of the sediment comprising these systems was delivered directly or indirectly by the river. The delta is the result of sediment deposition at and in the vicinity of the river mouth. The Chenier Plain was formed from sediment that was discharged from the river and transported to the west by currents and deposited along the western Louisiana coast and into easternmost Texas. Storms reworked the muddy sediments concentrating shells in long, low ridges called cheniers (see Fig. 5-30). The result is a low, muddy coastal region characterized by numerous elongate accretionary sediment bodies (Fig. 6-8).

The modern Mississippi Delta is a complex of several Holocene-aged sediment lobes (see Fig. 5-17). The currently active lobe in this complex extends only about 1000 years back in time. The others were abandoned as the main channel of the river avulsed multiple times over the past 7000 years. The result is a complex series of sediment accumulations that have produced extensive wetlands and very shallow subtidal environments which encompass many thousands of square kilometers.

This coast is very fragile due to its mud-dominated character and its low elevation, neither of which will fare well in the context of rising sea level. This coast is extremely important economically due to its petroleum production and also because of its natural environment for a diverse community of organisms. It is a nursery ground for many marine species and it is home to huge finfish and shellfish fisheries. Many mammals also pop-

Figure 6-8. Infrared photo showing accretion of sediments ultimately derived from the Mississippi Delta on the Bolivar Peninsula of easternmost Texas. These accretion bodies have since been eroded by hurricanes in 2005 to 2008. Photograph courtesy D. Nummedal.

ulate the delta including deer and nutria. The coast is also important as a protection for inland areas where there is extensive development and population.

Barrier Islands

The extensive wetland that comprises the Mississippi Delta is bound on the Gulf side by numerous low-lying barrier islands. These sand bodies are important and serve as protection for the more fragile wetlands landward. They are formed by the reworking of the distal end of the abandoned delta lobes. In general, these barriers rise less than 3 m above sea level and are vulnerable to beach erosion and washover by storms. The Chandelier Islands, the easternmost of the island chains, are natural and uninhabited by people. Most of this complex is less than a meter above sea level. To the west of the present mouth of the Mississippi River is another barrier complex that is larger and developed. Small communities and extensive petroleum installations occupy some of these areas (Fig. 6-9).

The moderate sea-level rise of 0.5 m will easily result in the destruction of the Chandeleurs and will cause major problems of erosion on

Figure 6-9. Petroleum industry installations impacting the wetlands in the Mississippi Delta. Photo courtesy Louisiana Geological Survey.

the larger, developed islands. Hurricanes Katrina and Ivan caused great destruction of the Chandelier Islands and storms such as these might destroy these islands before sea-level rise can do the job. The high sea-level rise scenario of a 1.0 m will destroy the other barrier islands as well. Even now, at the beginning of the 21st century, there are structures in place to prevent erosion on the developed islands. These structures have been in place a few years (Fig. 6-10) but their effectiveness is arguable. Some of the smaller and unoccupied barriers at the west end of this complex have already been nearly destroyed. It is likely that all of the barriers that surround the Mississippi Delta will be destroyed or rendered uninhabitable by the end of this century.

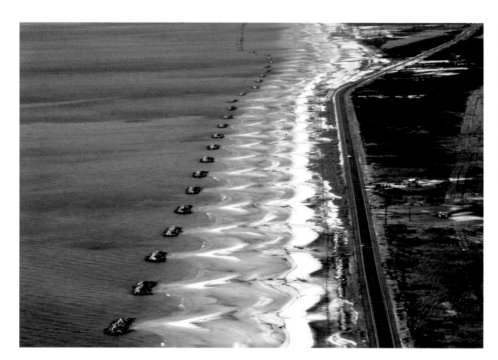

Figure 6-10. (top) Oblique aerial photo and (bottom) land-based closeup showing breakwaters placed to protect the Louisiana coast from erosion. Photographs courtesy G. Stone, Coastal Studies Institute, Louisiana State University.

Wetlands

The most vulnerable portion of the entire Gulf of Mexico coast is the wetland complex of the Mississippi Delta. The State of Louisiana, with the help of federal funding, is conducting a major effort to restore its coast. Committees have been established to address the problems and provide suggestions for solutions. A major product of these efforts is the special issue of the *Journal of Coastal Research* (1994) entitled "Scientific Assessment of Coastal Wetland Loss, Restoration, and Management in Louisiana." This region contains 40% of all of the wetlands in the United States. Currently, 65 km^2 of wetlands are being lost each year which is a decrease from the rate of loss in the 1960s through the 1980s. Since 1930, total wetland loss has been in excess of 4000 km^2, an area nearly twice the size of Rhode Island (see Fig 5-6).

Wetlands that extend for hundreds of kilometers are all at or very close to the same elevation. They occupy the upper portion of the intertidal zone. Along the Louisiana coast the mean tidal range is 40–50 cm. The upper portion of the intertidal zone where wetland vegetation grows is about 15–20 cm. With a rate of sea-level rise that is 1–2 cm/yr, it will take only a few years for many of these wetlands to be drowned. As a result, either the moderate or high sea-level rise scenario for the 21st century will destroy nearly all of the Louisiana coastal wetlands unless mitigation can be successful. The loss of wetlands along this coast is due to both human activities and to natural phenomena. The rate of sea-level rise in this region is also partly a result of both of these factors. Unfortunately, the causes of wetland loss due to sea-level rise or to human activities cannot be easily separated.

The global rate of sea-level rise is currently about 2 mm/yr. This is obviously only a small portion of the sea-level rise that is presently occurring on the Louisiana coast. Sediment compaction can be recognized and measured and it is a major component of the rate of relative sea-level rise. The combination compaction and eustatic rise, both natural phenomena, contribute most to the current rate of sea-level rise. The other major factor in sea-level rise that is typically considered is the withdrawal of fluids associated with petroleum extraction. However, many studies have shown that this factor is overestimated. Withdrawal of deep fluids has essentially no effect on surface subsidence. Only the fluid extraction from shallow wells or fields influences subsidence and that influence is only local. Unfortunately, the depth at which fluid extraction is or is not a contributor to local subsidence was not clearly defined by the studies. In addition, fluid withdrawals can activate faults in the subsurface and exacerbate subsidence and compaction.

Subsidence is a factor that can be overcome under proper conditions. If the Mississippi River provides enough sediment for wetland surface elevation to keep pace with subsidence and sea-level rise, then the wetlands can be maintained. There are several ways that human activity on the coast has prevented this from taking place. These include dams on the river, construction of levees, and dredging of canals. Each prevents sediment from reaching and nourishing these wetland environments.

The many navigational dams along the Mississippi River hold huge quantities of sediment that would normally be carried by the river. In the 19th century, a tremendous amount of land in the drainage basin was cleared and plowed as agriculture spread throughout the midwestern part of the country. In response, the modern delta grew rapidly from the time of the Civil War until the 1930s. Then, in the 1930s, dams began to be constructed on the river and shortly afterward oil and gas exploration began. Barge and ship traffic on the lower Mississippi River was heavy and the Corps of Engineers controlled the flow of the river. All of these factors contributed to problems on the delta itself. Coupled with the rise of sea level, these activities resulted in a general deterioration of the wetland environment.

As part of the control of flow and to aid in flood control, the Corps of Engineers diverted a significant portion of Mississippi River flow through the Atchafalaya River resulting in a new delta and important wetland construction (see Fig. 4-20). This has been a very positive factor in the maintenance of wetlands in this area. By contrast, the construction of levees along the entire lower Mississippi River has had a negative impact. These levees prevent overbank conditions and flooding in the interdistributary areas of the delta. As a result, sediment does not move over the delta surface to nourish the wetlands and help them maintain a proper elevation as sea level rises. Breeches in the levee system can produce large crevasse splays that become wetlands (Fig. 6-11). Unfortunately, the Corps repairs the breeches to maintain continuity of the levee system, eliminating continued nourishment of the splays with sediment. Another problem associated with the levee system is that it tends to cause the river

Figure 6-11. Large splay and the related crevasse where the levee is breeched on the main channel of the Mississippi River.

discharge to move into fairly deep water in a jet-like fashion. This causes most of the suspended sediment, that is composed of about 65% clay and 35% silt, to move beyond the delta. It is much of this sediment that makes its way to western Louisiana and becomes incorporated into the Chenier Plain. The remainder makes its way to deep water in the Gulf.

Numerous canals were dredged through the delta as part of the exploration and production of oil and gas. The spoil from the dredging was typically placed along the channel being excavated. These spoil banks behaved like levees and inhibited flux of sediment-laden water during spring high tides and especially during storm tides when a considerable amount of sediment is in suspension.

There is no question as to the causes of wetland loss on the Louisiana coast. It is also clear that if things continue as they have, the wetlands will mostly be gone by the end of the first 21st century. Given this scenario, what can be done?

First, it is obvious there is little that can be done to reduce the rate of subsidence due to compaction and the global rise in sea level. Fluid withdrawal does not appear to be a major factor so replacing fluids as others are taken will not provide a significant change. The only major factor that can be addressed is to increase the sediment that is being supplied to the delta. This will require major funding and also major changes in the management of the Mississippi River system. The key to this approach will be the actions of the Corps of Engineers with the support and cooperation with various environmental agencies at the state and federal level.

Mitigation Possibilities

As stated earlier, the barrier islands provide some protection for the wetlands of the delta plain. There appears to be little that can be done to mitigate the deterioration of these barriers. By the end of the 21st century they will be destroyed or reduced in area to the point that any human developments should be removed. The only possible recourse is construction of a huge seawall across the Gulf side of the islands. This will be very expensive, and will provide protection from storms but not from sea-level rise. Such a structure would not solve the problem but will provide more time to develop and carry out an evacuation plan for the developed islands.

Mitigation of wetland loss is predominantly related to providing sediment to permit maintenance of surface elevation relative to sea level. This means that the present flood control situations must be changed. This can be accomplished in different ways. First, the flood control practices of the Atchafalaya Basin need to be applied elsewhere. As much sediment-laden river discharge as possible should be directed to the Atchafalaya Basin and adjacent areas where wetlands can form. Another important factor is to cut crevasse-type openings in the levees with gates along the main channel to permit more flooding of the river discharge onto the delta plain. Opening the gates during high discharge events will generate splays that will provide the appropriate setting for establishing wetland vegetation. Both of these approaches to mitigating the current practices that are leading to wetland loss can be done with nominal cost and modest impact on human activities.

A more difficult approach would be to permit widespread flooding of the delta plain during periods of high discharge, typically during spring when meltwater and precipitation in the drainage basin lead to flooding conditions throughout much of the delta region. The only way that this can take place in a short time is for the Corps of Engineers to lower the elevation of the levees to permit flooding to occur. This obviously cannot take place upstream from New Orleans. However, using this approach below the city would lower levees over about 150 km of the lower river and would provide sediment over most of the modern delta lobe. This approach would also greatly diminish the loss of sediment to deep water via the main distributaries at South Pass, Southwest Pass, and North Pass (see Fig. 4-19).

Figure 6-12. Aerial photograph showing dredged canal with spoil on the delta plain of the Louisiana coast.

Another smaller scale but important effort is changing the canals that are widespread on the delta plain (Fig. 6-12). The spoil banks associated with these canals inhibit sediment-laden flood water from spreading over the wetland surface. Backfilling the canals would eliminate this problem. Another more expensive approach would be to spread the spoil across the present wetland to nourish it and compensate for the subsidence that is taking place.

These are idealistic approaches to the problems of the Louisiana coast because of costs, bureaucracy, and environmental considerations. It is unlikely that all of these approaches to mitigation of wetland loss will take place but any of them would be a major step toward addressing the problem.

Texas Coast

The Texas coast is dominated by barrier islands that enclose numerous estuaries and lagoons. There is considerable range in the present rate of subsidence from the northeast coast to the southwest coast at the border with Mexico. The part of this coast that borders Louisiana to and including the Houston-Galveston metropolitan area is experiencing the most rapid sea-level rise of more than 1 cm/yr locally. This area is also quite complicated in that there are several factors that contribute to this condition. South of this area on the Texas coast the rate of rise is only slightly higher that the global rate.

Sabine River Area

   This part of the northwestern Gulf Coast is dominated by thick, muddy sediments and is a region of oil and gas production. It is also an area where wetlands are widespread. This is an unfortunate combination and the result is that major subsidence occurred and wetlands were drowned in a short time (Fig. 6-13). Mitigating such conditions will require either a major reduction in fluid withdrawal and/or replacing these fluids on a volume for volume basis.

Figure 6-13. Maps and aerial photo of a wetland area near the Neches River in easternmost Texas showing the difference in the surface environments from 1956 to 1978. The heavy red lines are faults. From White, W. A. and R. A. Morton, 1997, Wetland losses related to fault movement and hydrocarbon production, southeastern Texas coast. Journal of Coastal Research 13:1305–320.

1956

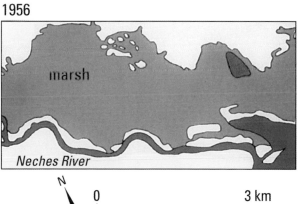

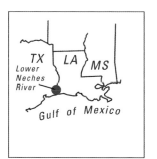

1978

Houston-Galveston area

Much of the cause for the high rate of sea-level rise in this area is attributed to subsidence caused by the combination of sediment compaction and fluid withdrawal. Thick sediment sequences associated with the ancestral Trinity River contribute to the compaction of the older sediments. The combination of withdrawals of groundwater for domestic and industrial uses and fluids associated with the petroleum industry are also important.

This area is second only to the Louisiana coast in terms of scrutiny of the causes and effects of rapid sea-level rise. Much of that has been summarized in a report by David Yoskowitz and James Gibeaut of Texas A&M University–Corpus Christi. Their predicted sea-level rise for this area over the 21st century is 0.7 m and 1.5 m for the moderate and extreme scenarios respectively. Because this region, in contrast to the Louisiana coast, is heavily populated, the effects of these sea-level rise scenarios will be different.

Many people will be displaced, and numerous buildings and water treatment plants will be impacted. The heavily industrialized areas (Fig. 6-14) associated with the petroleum industry, waste and superfund sites, and the Port of Houston will be greatly affected. Special considerations are necessary for the Galveston area. Hurricane Ike in 2008 showed that the area is quite vulnerable. A similar hurricane in future decades will be worse. There is now consideration of an 80 km long structure

Figure 6-14. Aerial photo showing industrial development adjacent to wetland in the Houston area. Photograph courtesy HDR Engineering.

Figure 6-15. (top) Flooding in the Galveston area and (bottom) destruction near Gilcrist, Texas on the Bolivar Peninsula associated with Hurricane Ike in September, 2008. Courtesy Wikipedia.

to protect Galveston Island and the adjacent bay area. Although such a structure would be very expensive and not aesthetically pleasing, this shows the kind of the thinking that is taking place about how to protect this important region.

Currently, development along the barrier coast in this area ranges from modest east of Galveston to very high in the city itself. There are few restrictions on the elevation or location of new construction. Several large residential developments have been built on the western half of Galveston Island that is low in elevation, much like the area of the city was prior to the 1900 hurricane. Hurricane Ike caused billions of dollars of damage in this area (Fig. 6-15). As sea level rises over this century even less intense hurricanes will cause similar results. These conditions are another reason for the interest in an extensive seawall along this barrier island.

The landward side of the barrier island and the Galveston Bay area are surrounded by saltmarsh wetlands. The rate of sea-level rise there is causing the same problems as those on the Mississippi Delta. Based on the model produced by Dr. Gibeaut for this area, 22% of the wetlands will be drowned by 2040. By the end of the century the low marsh wetland will be reduced by 77%.

Because this area is the most populated coastal area of those that are highly vulnerable to sea-level rise, the economic impact will be tremendous. The Houston area includes 18% of the population of Texas and much of it is vulnerable to rapidly increasing sea level. In their report, Yoskowitz and Gibeaut produced maps showing the effects of each of their sea-level scenarios (Fig. 6-16). In the Galveston area, 78% of households would be displaced under the 0.7 m sea-level rise scenario and 93% under the 1.5 m sea-level rise scenario. This would eliminate Galveston as a viable community on the Texas coast. There would also be significant impact under each scenario to water treatment facilities, waste disposal

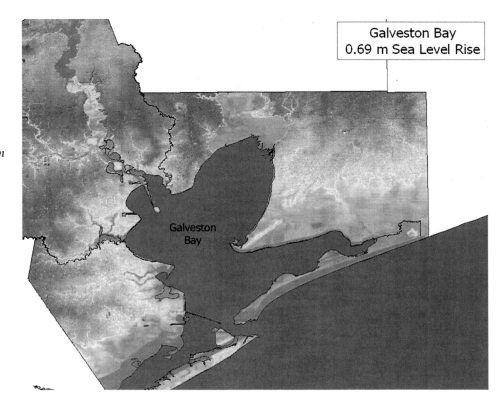

Figure 6-16. Map of the Galveston Bay area of Texas showing a hundred year storm surge on top of a sea level rise of 0.7 m through the 21ˢᵗ century. This map is based on water level during a 100-year storm surge. The light blue is the flooded area. Modified from Yoshkowitz, D. W. and J. C. Gibeaut, **The Socio-Economic Impact of Sea Level Rise in the Galveston Bay Region,** *Report For Environmental Defense Fund. Corpus Christi, Texas: Harte Research Institute for Gulf of Mexico Studies.*

areas, and superfund sites where toxic chemicals could be mobilized. The bottom line is that sea-level rise, with its concomitant direct and indirect effects, could wipe out the Galveston area and severely impact the Houston area by the end of the 21st century.

Texas Coast Barrier Islands

The Texas coast is bounded by a nearly continuous system of well-developed barrier islands except for the area near the mouth of the Brazos River. These barriers typically have well-developed dunes with elevations up to 10 m above sea level. Sea-level rise in this area is slightly above the eustatic rate. Hurricanes are common with varying consequences. There are only two significant areas of development along the barrier coast; the Coastal Bend area on the central coast and South Padre Island near the border with Mexico. As a result, the sea-level forecasts for the 21st century will have minimal influence on these parts of the coast.

Either sea-level rise scenario (0.5 m or 1.0 m) forecast by the Environmental Protection Agency will result in coastal retreat but the rate will be slow compared to the effect of hurricanes which impact this coast on nearly an annual basis. The two areas of development differ greatly. Development on central Mustang Island and northern Padre Island is primarily residential. Recent restrictions on development have established setback lines that are about 100 m from the beach and well landward of the dunes. This is a very forward looking approach to coastal development and is better than any other part of the Gulf Coast.

By contrast, the South Padre Island area is extensively developed with high rise hotels and condominiums and is a densely populated, well-established tourist destination. Setback restrictions are minimal and erosion problems on the beach have been common enough to require beach nourishment. Rising sea level over the 21st century will exacerbate the present situation.

The area of the Texas coast where barriers are absent is near and to the south of the mouth of the Brazos River (Fig. 6-17). Most of this coastal area is comprised of fluvial-deltaic sediments that compact to increase the rate of sea-level rise. Sand for development of barriers or mainland beaches is not available. The present seawall is designed to protect the Intracoastal Waterway. The projected sea-level rise scenarios for this

Figure 6-17. Aerial photo showing the mouth of the Brazos River where true barrier islands are absent. The delta here consists of barrier like sediment bodies that are separated from the mainland. Courtesy U.S. Department of Agriculture.

century will result in destruction of this structure and likely lead to the intersection of the shoreline with the ICW. The only way to prevent this from happening will be to add to the existing structure or relocate the ICW.

Coastal Bays

The numerous bays on the Texas coast range widely in size, shape, and landward discharge. Most have at least one sizeable river carrying sediment to the bays and depositing a bayhead delta. These deltas and the bay margins are the sites of extensive wetlands, mostly salt marshes. There are some areas where mangrove mangals also develop, generally comprised of black mangroves (*Avicennia germinans*). Only the bayhead deltas are currently receiving sediment to nourish the wetlands. The bay margins are sediment-starved. Both moderate and high projected sea-level rise rates will cause the saltmarsh environments to drown. The current practice of constructing marsh areas to mitigate sea-level rise and coastal construction practices are only temporary solutions. Similar to the Mississippi Delta area, there is only 10–20 cm of elevation over which the marsh vegetation can survive. During the 21st century, sea level will rise higher than these elevations and there are not many places where a transition from upland to wetland environments can take place. There is precedent for these changes as shown in subsurface data acquired and interpreted by the group from Rice University.

A coastal environment that is one of the most vulnerable to sea-level rise is wind-tidal flats. This environment is associated with areas that are little influenced by lunar tides, generally along the margins of backbarrier bays such as the Laguna Madre. The difference in elevation between the land side and the open water in the bay may be only a few centimeters. Movement of water by wind causes these flats to be inundated and exposed at irregular intervals. Wind-tidal flats look barren and unproductive, but are areas that are used extensively by migrating and wintering shorebirds and other aquatic birds. Rising sea level will drown this environment and eliminate it from the coast in a very short time, as little as a decade or two.

Some of these coastal bays have extensive industrial infrastructure on their margins. Examples are Nueces Bay and Corpus Christi Bay near Corpus Christi where there are seven oil refineries and a major bauxite processing plant, respectively. Further north on this coast at Lavaca Bay is another bauxite processing plant. These and other similar industrial complexes, along with related infrastructure, will be impacted by sea-level rise during this century. In addition, water quality in the adjacent estuaries will likely deteriorate.

Mexico Coast

There is little information on any details of what is predicted for the Mexican coast. This brief discussion will be based on inference and on comparison with similar areas on the U.S. Gulf Coast. With a couple of exceptions, this coast is not developed. The cities of Tampico, Veracruz, and Ciudad Carmen are the exceptions. All are heavily populated and depend somewhat on the coast for tourism and general economy. Coastal bays are elongate and parallel to the coast and the wetlands present are dominated by mangroves.

Predicted sea-level rise scenarios will not have a major impact on this coast except in the cities where adjustments in development will need to be made. Undoubtedly relocation of buildings, protection from rising water, and other modifications will be needed. Because the Mexican Gulf Coast is generally less developed and the rate of sea-level rise is near global rates, it is unlikely that the impacts of increasing sea level over this century will be as severe as those expected in many areas of the United States.

Island Coasts

Very little information is available for the coasts of Cuba and the Cayman Islands, both of which can be considered as part of the Gulf of Mexico. In the case of Cuba, the northwestern portion of that island is in the Gulf. This part of Cuba is geologically different from coasts in the rest of the Gulf. It is rocky with high relief, and will experience essentially no impact at either a 0.5 m or 1.0 m rise in sea level. In addition, the population is very sparse so the overall impact of sea-level rise on this part of the Gulf Coast during this century will be negligible.

By contrast, the Cayman Islands can expect major impact from either the moderate or high sea-level rise scenario. The low-lying islands are primarily tourist destinations and are fairly heavily developed, especially Grand Cayman Island. They are carbonate and have Pleistocene Stage 5e reefs (see Chapter 3) that protect the western side of the island and aeolianites (fossil dunes) on the eastern end. Most of the rest of the island is low and susceptible to erosion. Because of the extensive development on the west end of the island it is likely that various protective structures will need to be built to protect buildings and infrastructure. It is likely that sea-level rise will have a negative impact on the tourism industry on these islands.

Summary

The Gulf Coast will change markedly during the 21st century, largely as the direct or indirect result of rising sea level. Without question, some areas will be affected more than others. Changes in coastal management strategies will be necessary to maintain both built and natural environments. All of this will be very expensive. All levels of government, from state and local to federal, will be involved and will be required to make decisions and spend large amounts of money.

The most complicated and severely impacted region is the Louisiana coast. This vulnerable coast is not only susceptible to sea-level rise, but also to hurricanes. Is their enough money available to save it? That is a question that must be answered in the very near future. The other major problem area is the Houston-Galveston area where a huge population and extensive industrial complexes are vulnerable to sea-level rise and its consequences. These situations need to be addressed and major decisions must be made to deal with the rapidly rising sea level in that area.

While arguments continue about the causes of global warming, sea level is rising. About that, there is no question. The cause of sea level is immaterial. The fact is that it is happening. Sea level is rising and at the rate of that rise is increasing. It is time to do something about it!

Important Readings

Flick, R. E., K. Gooderham, L. Benedet and T. Campbell, (eds.). 2009. *Sea Level Rise—Facing our Future.* Shore and Beach Special Issue 77. Ft Myers, Fla.: American Shore and Beach Preservation Association. A collection of papers on both global and regional sea-level situations with forecasts and recommendations for mitigation.

Hoffman, J. S., D. Keyes, and J. G. Titus. 1983. *Projecting Future Sea Level Rise: Methodology, Estimates to the Year 2100 and Research Needs.* Washington, D.C.: U.S. Environmental Protection Agency. This is an older version of the attempts to predict sea-level changes. Now the IPCC approach is attempting to do this near-impossible task.

IPCC (Intergovernmental Panel on Climate Change). 2007. *IPCC Fourth Assessment Report: Climate Change 2007,* 4 vol. Cambridge, UK: Cambridge University Press. The most current information and data from the Intergovernmental Panel on Climate Change that is available on their website.

Mooers, C. N. K., (ed.). 1999. *Coastal Ocean Prediction.* Coastal and Estuarine Studies No. 56. Washington, D.C.: American Geophysical Union. Another attempt at predicting what will happen in the future of sea-level change.

National Research Council (NRC). 1987. *Responding to Changes in Sea Level.* Washington, D.C.: National Academy Press. A panel of experts discusses their opinions on the future of sea level.

National Research Council (NRC). 1994. *Environmental Science in the Coastal Zone: Issues for Future Research.* Washington, D.C.: National Academy Press. This is, in a way, an update on the volume published by the Academy in 1987.

Thieler, E. F. and E. S. Hammar-Klose. 2000. *National Assessment of Coastal Vulnerability to Future Sea-Level Rise: Preliminary Results for the U.S. Gulf of Mexico Coast.* USGS Open-File Report 00–179. Washington, D.C.: U.S. Department of Interior, U.S. Geological Survey. A systematic and objective examination of coastal vulnerability on the Gulf Coast of the United States.

Titus, J. G. and J. V. K. Narayanan. 1995. *The Probability of Sea Level Rise.* Washington, D.C.: U.S. Environmental Protection Agency. The US-EPA attempt to predict the sea-level future.

Yoskowitz, D. W., J. C. Gibeaut, and A. McKenzie. 2009. *The Socio-Economic Impact of Sea Level Rise in the Galveston Bay Region.* Report to the Environmental Defense Fund. Corpus Christi, Tex: Harte Research Institute for Gulf of Mexico Studies. An important, site-specific study of the sea-level situation in one of the most populated metropolitan areas of the Gulf Coast.

Index